Recursive Mind

Understanding the Brain's Mirroring
Model and Pattern Synergy

By

Behzad Ghorbani

Copyright © 2024 by Behzad Ghorbani

All rights reserved. No part of this publication may be reproduced, distributed, or transmitted in any form or by any means, including photocopying, recording, or other electronic or mechanical methods, without the prior written permission of the author, except in the case of brief quotations used in critical reviews or other non-commercial contexts permitted by law.

This work was conceived and authored by Behzad Ghorbani. The creative process employed the Hyper Hybrid Intellect model, an advanced methodological synthesis integrating human reflection with digital refinement. This model functions as a recursive mirror of thought, enabling the structural and analytical optimisation of complex theoretical material. It enhances clarity and multidimensional reasoning through *Comparative Matrix Analysis*, *Mirroring Hemisphere Fractal Analysis*, *Totalisation of Details*, and the *Recursive Realism Perspective*, ensuring a uniquely crafted and rigorously developed narrative.

1st Edition

Published by Amazon.com

This book is independently published by the author through Amazon Kindle Direct Publishing (KDP).

10 9 8 7 6 5 4 3 2

October 2024

London

Paperback: ISBN: 9798343010039
Hardcover: ISBN: 9798345270745

CONTENTS

	Preface	6
1	Introduction to Recursive Mirroring and Pattern Formation	11
2	The Brain's Symmetry and Neural Architecture	18
3	Pattern Recognition as the Basis of Human Behaviour	24
4	Creativity and Intellectual Synergy	32
5	Emotions, Memory, and Pattern Integration	40
6	Biological Analogies in Pattern Formation	49
7	The Brain's Role in Stereotyping and Survival	58
8	Memory, Learning, and Neuroplasticity	67
9	The Role of Recursive Patterns in Language and Literature	76

10	Music and the Synergy of Patterns	86
11	Art, Architecture, and the Brain's Visual Patterns	95
12	Synergy in Social Behaviour and Group Dynamics	104
13	Recursive Patterns in Moral Decision-Making and Ethical Behaviour	114
14	Recursive Thinking in Creativity, Science, and Innovation	123
15	Recursive Patterns in Time, Memory, and the Self	132
16	Recursive Patterns in Consciousness and Self-Awareness	141
17	Recursive Processes in Learning and Education	150
18	Recursive Patterns in Human Relationships and Social Dynamics	159
19	Recursive Processes in Cultural Evolution and Societal Change	168
20	Recursive Processes in Global Systems and Interconnected Societies	177
21	Synthesising Recursive Patterns: Complexity, Adaptation, and Human Progress	186
	Bibliography	195

Further Reading 202

Preface

The human brain is, without a doubt, the most intricate and powerful system we know of in the universe. Yet, despite our ever-expanding knowledge in neuroscience, psychology, and cognitive science, the full scope of how it works, how it generates thoughts, creativity, consciousness, and complex behaviours, remains an enigma. This book presents a model and hypothesis built on the foundation of recursive processes, offering a new way to understand the brain's mechanisms and, more importantly, their far-reaching implications in shaping everything from personal cognition to societal structures and global systems.

The recursive model proposed in this work is grounded in the idea that the brain functions by mirroring and reflecting patterns repeatedly. At the heart of this hypothesis is the notion that the left and right hemispheres of the brain are constantly engaged in recursive mirroring of each other's processes, and this dynamic interplay results in the generation of everything from conscious thought to complex problem-solving. This mirroring process is not limited to the brain's hemispheres alone but extends to the way we process external stimuli and internalise emotions, behaviours, and abstract concepts.

What makes this model unique is that it frames the brain as a recursive system, a dynamic interplay of patterns that continuously evolve through reflection and refinement. It is a model that, once understood, offers profound implications not only for how we think about the brain itself but also for how we understand human behaviour, creativity, and learning. More importantly, it provides a framework that bridges disciplines, from neuroscience and psychology to social theory and cultural evolution, offering a unified explanation for the ways individuals and societies adapt and evolve over time.

The advantages of this recursive model are numerous and multifaceted. First, it introduces a new lens for understanding cognition and learning. The recursive nature of thought processes allows us to refine and build on previous knowledge, mirroring the same dynamic feedback loops that we see in adaptive systems across nature. By conceptualising learning as a recursive loop, where the brain mirrors patterns, reflects on them, and refines them, this model helps to explain how we acquire new skills, develop expertise, and adapt to novel challenges.

Second, the recursive model offers an explanation for the brain's role in generating creativity. Whether it is in the creation of art, music, literature, or scientific innovation, recursion allows for the dynamic layering of ideas, where patterns are mirrored and dislocated, resulting in new forms of expression. This model sheds light on how individuals like Pablo Picasso, Ludwig van Beethoven, and Salman

Rushdie were able to generate deeply innovative and resonant works through recursive processes that allowed them to maintain, evolve, and transform motifs and themes over time.

Third, the model has profound implications for understanding social behaviour and group dynamics. Human beings are inherently social creatures, and much of our behaviour is driven by the mirroring and refinement of social patterns. The recursive model explains how empathy, collaboration, and even conflict arise from recursive loops of social interaction, where individuals reflect and adapt their behaviours in response to others. It also helps clarify how cultural evolution, the process by which societies transmit, refine, and innovate their traditions and beliefs, mirrors recursive biological and cognitive processes.

Finally, this model offers a robust framework for addressing global systems and societal change. In an increasingly interconnected world, understanding how recursive processes shape everything from economic systems to environmental policies is critical. The recursive loops that connect local actions to global consequences highlight the importance of adaptive thinking and collaborative problem-solving. By framing global systems as recursively interconnected, this model offers new pathways for addressing pressing challenges such as climate change, technological disruption, and social inequality.

While this book delves into the technical aspects of recursion in the brain and behaviour, it also provides concrete examples and case studies to illustrate how these recursive processes play out in real-world scenarios. From the musical compositions of Beethoven to the cultural shifts driven by technological innovation, the recursive model explains the underlying mechanisms of change, adaptation, and innovation in both individuals and societies.

This hypothesis, built on the recursive functioning of the brain and its broader applications, opens a new frontier in understanding the most fundamental aspects of human existence, how we think, how we create, how we relate to one another, and how we evolve. It is my hope that the model presented here will serve not only as a tool for understanding the complexity of human cognition and behaviour but also as a guide for those seeking to apply these insights to education, social policy, and the ongoing challenges of the 21st century.

Behzad Ghorbani

London, September 2024

Chapter 1:

Introduction to Recursive Mirroring and Pattern Formation

The human brain is a remarkable organ, not only in terms of its biological structure but also in its cognitive capabilities. The central hypothesis of this book is that much of human cognition, behaviour, and creativity can be understood through two closely related processes; recursive mirroring and pattern formation. Acting together, these processes enable the brain to generate complex behaviour, solve problems, and recognise structured regularities within both internal experience and the external environment.

Recursive mirroring refers to the brain's capacity to reflect and reprocess information across interacting neural systems, particularly between the two cerebral hemispheres, creating an ongoing loop of exchange and refinement. This mirroring is not a simple repetition of signals but involves continuous feedback through which patterns are progressively stabilised, modified, or reorganised. Pattern formation, by contrast, concerns the brain's ability to recognise, generate, and retain structured configurations derived from sensory input, memory, and experience. These configurations function as the operative units of thought, behaviour, and perception.

Taken together, these processes offer a model of how the brain integrates information by forming structured patterns and reprocessing them across distributed regions to refine interpretation and generate new insights. Rather than treating hemispheric specialisation as a division of labour between isolated systems, this perspective considers the brain as a unified network in which information is continually reworked through recursive feedback.

Understanding how patterns are generated is therefore essential. Neural clusters form microcircuits capable of adopting states of activation or inactivation. These circuits are not pre-organised into fixed representations but remain available for dynamic association as information is processed. When sensory input or cognitive demand arises, these circuits become activated in specific configurations, producing patterns that can subsequently be re-engaged and refined through recursive interaction.

Within this framework, the cerebral cortex may be understood as a system capable of reflecting and reorganising patterns through bilateral interaction. The left and right hemispheres, connected through the corpus callosum, allow information processed in one region to be re-presented to another. This recursive exchange does not terminate after a single cycle but continues across multiple passes, each introducing minor variations or dislocations in the original configuration. These dislocations are not errors but adaptive modifications that enable flexibility and creative adjustment in response to changing demands.

Successive cycles of reflection generate layered configurations that may align or overlap to produce structures ranging from broad cognitive strategies to subtle emotional or perceptual responses. It is through these incremental variations that the brain adapts to new information, generates novel associations, and responds effectively to environmental change. In this sense, adaptability and creativity arise from the capacity to refine patterns through repeated interaction.

The implications of this model extend beyond a descriptive account of neural processing. They offer a basis for understanding creativity, intuition, and complex problem-solving as outcomes of iterative refinement rather than isolated moments of inspiration. Mirror writing, as demonstrated by Leonardo da Vinci, may be regarded as an explicit behavioural expression of this reflective capacity. Although rarely developed consciously, the underlying potential for such mirroring exists within every brain and may be strengthened through practice.

Creativity and innovation can therefore be viewed as the cumulative result of recursive processes that preserve, reorganise, and layer patterns until new connections emerge. This perspective helps explain why artists, thinkers, and scientists often maintain recognisable motifs across seemingly unrelated work; patterns refined in one domain may be re-engaged and adapted in another, producing continuity within innovation.

This approach also informs our understanding of abstract concepts. Ideas such as success or failure, love or loss, are not retained as isolated facts but as integrated patterns combining sensory experience, emotion, and cognitive association. Reflecting upon such concepts involves the reactivation and modification of these patterns, allowing meaning to deepen over time. The same recursive engagement underlies our response to art, literature, or music, where perceived patterns may evoke layered emotional and experiential associations.

This chapter establishes the conceptual basis for examining how recursive mirroring and pattern recognition contribute not only to individual cognition and behaviour but also to creativity, social interaction, and intellectual development. Subsequent chapters will explore how these processes operate across different domains, providing a unified perspective on the ways in which the brain generates, interprets, and adapts patterned experience.

Reference List

Anderson, J.R., 2015. *Cognitive Psychology and Its Implications*. 8th ed. New York: Worth Publishers.

Beaty, R.E., Benedek, M., Silvia, P.J. and Schacter, D.L., 2016. Creative cognition and brain network dynamics. *Trends in Cognitive Sciences*, 20(2), pp.87–95.

Clark, A., 2013. Whatever next? Predictive brains, situated agents, and the future of cognitive science. *Behavioural and Brain Sciences*, 36(3), pp.181–204.

Dehaene, S., 2014. *Consciousness and the Brain: Deciphering How the Brain Codes Our Thoughts*. New York: Viking Press.

Damasio, A., 2010. *Self Comes to Mind: Constructing the Conscious Brain*. London: Vintage.

Edelman, G.M., 2006. *Second Nature: Brain Science and Human Knowledge*. New Haven: Yale University Press.

Friston, K., 2010. The free-energy principle: A unified brain theory? *Nature Reviews Neuroscience*, 11(2), pp.127–138.

Gazzaniga, M.S., 2008. *Human: The Science Behind What Makes Us Unique*. New York: HarperCollins.

Hebb, D.O., 1949. *The Organisation of Behaviour: A Neuropsychological Theory*. New York: Wiley.

Hofstadter, D.R., 2007. *I Am a Strange Loop*. New York: Basic Books.

Kandel, E.R., 2012. *The Age of Insight: The Quest to Understand the Unconscious in Art, Mind, and Brain*. New York: Random House.

Kolb, B. and Whishaw, I.Q., 2015. *Fundamentals of Human Neuropsychology*. 7th ed. New York: Worth Publishers.

Lakoff, G. and Johnson, M., 1999. *Philosophy in the Flesh: The Embodied Mind and Its Challenge to Western Thought*. New York: Basic Books.

McGilchrist, I., 2009. *The Master and His Emissary: The Divided Brain and the Making of the Western World*. London: Yale University Press.

Merzenich, M.M., 2013. *Soft-Wired: How the New Science of Brain Plasticity Can Change Your Life*. San Francisco: Parnassus Publishing.

Norman, D.A., 2013. *The Design of Everyday Things*. Revised ed. New York: Basic Books.

Ramachandran, V.S., 2011. *The Tell-Tale Brain: A Neuroscientist's Quest for What Makes Us Human*. London: Vintage.

Schacter, D.L., 2012. *Searching for Memory: The Brain, the Mind, and the Past*. New York: Basic Books.

Tononi, G., 2008. Consciousness as integrated information: A provisional manifesto. *Biological Bulletin*, 215(3), pp.216–242.

Ward, T.B., Smith, S.M. and Finke, R.A., 1999. *Creative Cognition*. Cambridge, MA: MIT Press.

Chapter 2:

The Brain's Symmetry and Neural Architecture

To understand the processes of recursive mirroring and pattern formation described in the previous chapter, it is necessary to examine the structural organisation that supports these operations. The brain is not an undifferentiated mass of neural tissue but a highly organised system characterised by both symmetry and functional specialisation. Its capacity to engage in recursive mirroring arises from its bilateral architecture; the division into left and right hemispheres, each contributing distinct yet interconnected roles in cognition, behaviour, and perception.

Central to this symmetry is the corpus callosum, a dense bundle of nerve fibres linking the two hemispheres. This structure allows for continuous communication between the left and right sides of the brain, enabling information processed in one region to be re-presented to the other. In doing so, it functions not merely as a passive connection but as an active pathway through which neural patterns can be exchanged, re-engaged, and refined through recursive interaction.

Although the hemispheres are structurally similar, they are not functionally identical. The left hemisphere has traditionally been associated with analytical processing, language, and sequential reasoning. It tends to decompose tasks into component elements and analyse information in a structured, linear manner. The right hemisphere is more commonly linked with spatial awareness, contextual integration, and holistic processing. It supports the synthesis of information across multiple domains, allowing patterns to be perceived as coherent wholes rather than as isolated components.

These functional distinctions operate within a continuously interacting system. Analytical processes occurring in one hemisphere may be reinterpreted through pattern integration in the other, and vice versa. Through this reciprocal exchange, the brain is able to move beyond basic problem-solving toward abstract reasoning, creativity, and intuitive judgement. Recursive interaction between hemispheres enables the refinement of emerging patterns, allowing them to be stabilised, modified, or reorganised in response to changing cognitive demands.

This architecture also supports the emergence of pattern dislocation, a feature central to the generation of novel ideas. As information is reflected across interacting neural systems, slight variations may arise in the configuration of activity. These variations are not errors but adaptive modifications that introduce flexibility into the brain's operations. In a manner analogous to biological evolution, where small mutations allow for adaptive change, these neural dislocations permit the recombination and extension of existing patterns, supporting creative thought and behavioural innovation.

At a more fundamental level, these processes occur through networks of neurons organised into microcircuits. Neurons rarely function in isolation; they form interconnected assemblies capable of representing states of activation and inactivation. Such microcircuits may be regarded as the operative units through which cognitive patterns are formed. When billions of these units become active in coordinated configurations, they generate the structured patterns that underlie perception, memory, and behaviour.

These microcircuits are distributed throughout the brain in a manner that supports flexibility rather than rigid localisation. They are not confined to predetermined functional roles but may be dynamically organised according to task demands. As a result, the brain continually forms and re-forms functional networks in response to new information and experience.

The cerebral cortex plays a particularly important role in this process. As the outer layer of the brain responsible for higher-order functions such as perception, decision-making, and planning, it integrates activity across multiple regions. Its lobular organisation reflects broad functional tendencies; the frontal lobe contributes to planning and problem-solving, the parietal lobe to sensory integration and spatial reasoning, the temporal lobe to auditory processing and language, and the occipital lobe to visual perception.

Despite these regional specialisations, cortical activity remains highly integrated. Complex tasks, whether mathematical reasoning or musical composition, require coordinated engagement across multiple areas. Information may be analysed in one region, contextualised in another, and linked to memory in a third. Through recursive feedback between these regions, diverse forms of information such as sensory input, emotional context, and prior experience may be synthesised into coherent responses.

Such recursive feedback loops are fundamental to the brain's capacity for pattern recognition and adaptive learning. When confronted with a problem, the brain may initially engage frontal regions to analyse its structure, subsequently reprocess this analysis through parietal systems for spatial or relational context, and then draw upon temporal regions for relevant memory associations. The iterative exchange of information across these regions allows a progressively refined interpretation to emerge.

Examining the brain's neural architecture in this way provides insight into how structured patterns are formed and modified. Recursive interaction across distributed systems supports not only the recognition of existing patterns but also the generation of new ones. Through this process, the brain is able to learn from experience, adapt to environmental change, and develop increasingly sophisticated modes of thought.

This structural perspective serves as a bridge to the cognitive processes explored in subsequent chapters. The capacity to form, reflect, and refine patterns is not a static feature but an ongoing dynamic activity underlying perception, reasoning, and creativity. Understanding this architecture allows for a deeper appreciation of how recursive interaction contributes to both individual cognition and the broader patterns of behaviour that shape human experience.

References

Bear, M.F., Connors, B.W. and Paradiso, M.A., 2020. *Neuroscience: Exploring the Brain*. 4th ed. Philadelphia: Wolters Kluwer.

Blakemore, S.J. and Frith, U., 2005. *The Learning Brain: Lessons for Education*. Oxford: Blackwell Publishing.

Bloom, F.E., Lazerson, A. and Hofstadter, L., 2001. *Brain, Mind, and Behaviour*. 3rd ed. New York: Worth Publishers.

Gazzaniga, M.S., Ivry, R.B. and Mangun, G.R., 2019. *Cognitive Neuroscience: The Biology of the Mind*. 5th ed. New York: W.W. Norton.

Geschwind, N., 1965. Disconnexion syndromes in animals and man. *Brain*, 88(2), pp.237–294.

Kolb, B. and Whishaw, I.Q., 2015. *Fundamentals of Human Neuropsychology*. 7th ed. New York: Worth Publishers.

Mesulam, M.M., 2000. *Principles of Behavioural and Cognitive Neurology*. 2nd ed. Oxford: Oxford University Press.

Mountcastle, V.B., 1997. The columnar organization of the neocortex. *Brain*, 120(4), pp.701–722.

Purves, D., Augustine, G.J., Fitzpatrick, D., Hall, W.C., LaMantia, A.S. and White, L.E., 2018. *Neuroscience*. 6th ed. Sunderland, MA: Sinauer Associates.

Ramachandran, V.S., 2011. *The Tell-Tale Brain: A Neuroscientist's Quest for What Makes Us Human*. London: Vintage.

Sperry, R.W., 1968. Hemisphere deconnection and unity in conscious awareness. *American Psychologist*, 23(10), pp.723–733.

Tononi, G. and Koch, C., 2015. Consciousness: Here, there and everywhere? *Philosophical Transactions of the Royal Society B*, 370(1668), pp.1–18.

Zaidel, E. and Iacoboni, M., 2003. *The Parallel Brain: The Cognitive Neuroscience of the Corpus Callosum*. Cambridge, MA: MIT Press.

Chapter 3:

Pattern Recognition as the Basis of Human Behaviour

The ability to recognise patterns is one of the most fundamental aspects of human cognition. It is through pattern recognition that we make sense of the world around us, navigate social interactions, solve problems, and anticipate the future. In essence, pattern recognition forms the backbone of all human behaviour, from the most basic reflexive actions to the most complex decision-making processes. It allows us to process sensory information, recall memories, and generate appropriate responses, all of which are vital to survival and adaptation within a constantly changing environment.

Within the framework of the recursive mirroring model, pattern recognition is not a static or one-time event but an ongoing dynamic operation. The brain continuously mirrors and refines the patterns it perceives, allowing for the integration of new information and the continual adjustment of behaviour. This recursive mirroring occurs both across hemispheres and within distributed neural networks, enabling the brain to detect and respond to patterns at multiple levels of complexity in real time. Patterns are not simply observed; they are repeatedly engaged, compared with previous configurations, and modified as new information becomes available.

When we encounter a familiar situation, the brain rapidly matches incoming stimuli, whether visual, auditory, tactile, or emotional, to previously established configurations derived from past experience. This process of matching and retrieving patterns from memory allows us to interpret the situation and generate an appropriate response without conscious deliberation. For example, when we see a red traffic light, the brain immediately recognises the pattern associated with stopping and triggers the relevant motor response. In this case, behaviour relies upon a highly stabilised pattern that has been reinforced through repetition over time.

However, not all patterns are simple or immediately recognisable. In more complex situations, such as navigating social interactions or solving novel problems, the brain engages in multiple layers of recursive pattern recognition. It mirrors perceived patterns against prior experience, introduces new contextual elements, and refines its interpretation with each iteration. This recursive process allows the brain to recognise subtle cues, anticipate potential outcomes, and generate flexible responses that are appropriate to the demands of the situation.

Consider the act of conversation. On the surface, conversation may appear to be a straightforward exchange of words, yet beneath this exchange lies a dense network of pattern recognition. The brain simultaneously interprets the speaker's tone, facial expressions, gestures, and posture, as well as the broader social context in which the interaction occurs. Emotional undertones, even when not explicitly expressed, are incorporated into this interpretation. Each of these cues is mirrored against existing social knowledge and prior experience. Through recursive engagement, the brain refines its understanding in real time, allowing adjustments in response as the interaction unfolds.

Predictive processes play a crucial role in this activity. The brain does not merely respond to patterns after they occur but anticipates likely developments based on previously recognised configurations. During conversation, for instance, the brain continually predicts how the speaker might respond, how the topic might shift, and which emotional cues may emerge next. When predictions align with unfolding events, the existing pattern is reinforced. When they do not, the pattern is modified, introducing a slight variation or dislocation that refines future expectations.

This cycle of anticipation and adjustment is equally evident in problem-solving and decision-making. When confronted with a challenge, the brain draws upon patterns derived from similar past experiences, mirrors these patterns against the present context, and tests possible responses. Each potential solution is recursively evaluated and refined until a response emerges that satisfies both logical requirements and emotional considerations.

The flexibility of this pattern recognition system is crucial for survival in environments that are rarely predictable. Unlike a machine operating under fixed instructions, the human brain is capable of adapting its patterns in real time. This adaptability is particularly important in social behaviour, where interactions vary from moment to moment. The capacity to recognise social patterns and adjust them according to context allows individuals to navigate relationships with a degree of fluidity that would not be possible through rigid behavioural rules.

A striking example of this adaptability can be observed in unregulated traffic environments. In situations where formal rules, signals, or lane markings are absent, drivers must rely on their ability to recognise patterns in the movement of others. They assess the speed and trajectory of vehicles, the proximity of pedestrians, and the overall density of traffic. By mirroring these patterns against prior driving experience, they generate split-second decisions that allow traffic to flow despite the absence of explicit coordination. Through instinctive cooperation, individual actions become synchronised, forming a self-organising system in which behaviour is guided by shared recognition of movement patterns rather than formal regulation.

In such circumstances, the brain is not only recognising patterns but actively generating new ones. Each adjustment in response to changing traffic conditions refines the driver's internal representation of the situation, enabling flexible and adaptive behaviour. Similar processes operate in everyday decision-making. Whether choosing clothing, responding to correspondence, or organising daily tasks, the brain continually recognises patterns in both the environment and internal states. These patterns are recursively mirrored and refined according to goals, past experience, and contextual demands, ensuring that behaviour remains efficient and appropriate.

Automatic behaviour represents another outcome of stabilised pattern recognition. Actions such as walking, brushing one's teeth, or navigating familiar routes are governed by patterns that have been refined through repeated use. Once stabilised, these patterns require minimal conscious oversight and may be executed with little cognitive effort. Nevertheless, flexibility remains possible. When conditions change, such as when a familiar route is blocked, the brain introduces a dislocation that allows for an alternative course of action.

Spontaneous behaviour also emerges from recursive pattern recognition. When confronted with unexpected events, the brain draws upon previously established configurations to generate responses without prolonged deliberation. These responses are not random but arise from the rapid mirroring of past experience against present circumstances, allowing for behaviour that is both immediate and contextually appropriate.

In this way, spontaneous action may be understood as the expression of recursive pattern engagement across cognitive and emotional domains. The capacity to synchronise these domains enables individuals to respond to novelty with creativity, intuition, and intelligence.

In summary, pattern recognition underlies all aspects of human behaviour. Through recursive mirroring, the brain continually refines the patterns upon which perception, decision-making, and interaction depend. This ongoing refinement ensures that behaviour remains adaptive in the face of changing circumstances. As subsequent chapters will demonstrate, these processes extend beyond immediate action to influence creativity, intellectual synergy, and emotional experience, shaping not only behaviour but also the broader patterns of thought through which individuals engage with the world.

References

Barsalou, L.W., 2008. Grounded cognition. *Annual Review of Psychology*, 59, pp.617–645.

Clark, A., 2016. *Surfing Uncertainty: Prediction, Action, and the Embodied Mind*. Oxford: Oxford University Press.

Friston, K., 2010. The free-energy principle: A unified brain theory? *Nature Reviews Neuroscience*, 11(2), pp.127–138.

Gigerenzer, G. and Todd, P.M., 1999. *Simple Heuristics That Make Us Smart*. Oxford: Oxford University Press.

Goldstone, R.L., 1998. Perceptual learning. *Annual Review of Psychology*, 49(1), pp.585–612.

Kahneman, D., 2011. *Thinking, Fast and Slow*. London: Penguin Books.

Klein, G., 1998. *Sources of Power: How People Make Decisions*. Cambridge, MA: MIT Press.

Neisser, U., 1967. *Cognitive Psychology*. New York: Appleton-Century-Crofts.

Norman, D.A. and Shallice, T., 1986. Attention to action: Willed and automatic control of behaviour. In: R.J. Davidson, G.E. Schwartz and D. Shapiro, eds. *Consciousness and Self-Regulation*. New York: Plenum Press, pp.1–18.

Rumelhart, D.E. and McClelland, J.L., 1986. *Parallel Distributed Processing: Explorations in the Microstructure of Cognition*. Cambridge, MA: MIT Press.

Schank, R.C. and Abelson, R.P., 1977. *Scripts, Plans, Goals and Understanding*. Hillsdale, NJ: Lawrence Erlbaum Associates.

Simon, H.A., 1996. *The Sciences of the Artificial*. 3rd ed. Cambridge, MA: MIT Press.

Sun, R., 2002. *Duality of the Mind: A Bottom-Up Approach Toward Cognition*. Mahwah, NJ: Lawrence Erlbaum Associates.

Thorndike, E.L., 1898. Animal intelligence: An experimental study of the associative processes in animals. *Psychological Review Monograph Supplements*, 2(4), pp.1–109.

Tolman, E.C., 1948. Cognitive maps in rats and men. *Psychological Review*, 55(4), pp.189–208.

Chapter 4:

Creativity and Intellectual Synergy

Creativity, often regarded as a mysterious and uniquely human ability, is deeply rooted in the processes of recursive mirroring and pattern recognition described in the preceding chapters. Rather than emerging as a sudden spark of inspiration or an isolated moment of insight, creativity may be understood as the outcome of a sustained recursive process in which the brain mirrors, refines, and reorganises patterns over time, allowing for the emergence of new ideas and adaptive solutions. Intellectual synergy, the integration of distinct cognitive processes into a coherent and effective whole, arises from the same underlying principles. When the brain's hemispheres and distributed neural regions exchange and re-engage information through recursive interaction, creative breakthroughs and deeper intellectual insights may emerge.

In this chapter, we will examine how creativity and intellectual synergy arise from these recursive processes. Particular attention will be given to the role of pattern dislocation, minor variations introduced within repeated configurations, which provide the basis for novelty while maintaining coherence with prior experience. These variations enable the brain to generate ideas that are not only original but also meaningful within existing cognitive and cultural frameworks. Illustrative examples may be found in the work of individuals such as Leonardo da Vinci, Pablo Picasso, and Ludwig van Beethoven, as well as within broader domains including art, music, and scientific discovery.

At the core of creative activity lies the brain's capacity to recognise established patterns and subtly modify them. These modifications are not random errors but adaptive variations introduced during recursive engagement with existing configurations. When a composer develops a musical theme, for instance, previously encountered harmonic and rhythmic structures are mirrored against new compositional intentions. Slight alterations in tone, timing, or sequence introduce variations that test the stability of familiar patterns while preserving their recognisability. Through successive refinements, these variations may give rise to entirely new harmonic relationships or structural innovations.

Beethoven's work provides a notable example of this process. Across his career, he continually experimented with form and thematic development. In his later symphonies, musical motifs are introduced, reflected, and progressively transformed through recursive engagement. The resulting compositions maintain structural coherence while incorporating dramatic shifts in tone, rhythm, and harmony. This balance between continuity and innovation illustrates the brain's ability to refine patterns recursively while allowing for meaningful variation.

A comparable process may be observed in the visual arts. Pablo Picasso maintained recurring motifs across his artistic development, yet introduced variations that dislocated familiar visual patterns. His transition from classical realism to Cubism demonstrates how established perceptual configurations may be mirrored, fragmented, and recombined to produce novel modes of representation. Cubism itself may be understood as an exploration of how visual patterns can be reorganised through recursive modification, allowing perception to extend beyond conventional spatial frameworks.

Creative integration is not confined to artistic domains. Leonardo da Vinci exemplifies the synergy between analytical reasoning and imaginative exploration. His sketches of flying machines, derived from observations of avian movement, reflect the mirroring of natural patterns combined with deliberate modification. By recursively engaging with these patterns, he generated technological concepts that extended beyond direct imitation. His use of mirror writing may be regarded as a behavioural expression of reflective processing, symbolising the iterative refinement of patterns across domains of inquiry.

These examples illustrate the principle of intellectual synergy, wherein distinct modes of cognition, logical, perceptual, and emotional, are integrated into a unified system. Such synergy emerges not from the dominance of a single cognitive style but from the brain's capacity to mirror and refine patterns across multiple domains simultaneously. Through recursive engagement, abstract concepts and sensory experiences may be combined, allowing analytical reasoning to interact with intuitive judgement.

Scientific discovery offers further evidence of this integrative process. Many significant breakthroughs have resulted from the synthesis of diverse forms of knowledge rather than from purely linear reasoning. Albert Einstein's development of the theory of relativity, for example, involved imaginative thought experiments alongside formal mathematical analysis. By envisioning himself travelling alongside a beam of light, Einstein engaged in recursive patterning that allowed abstract relationships between space and time to be explored conceptually before being formalised mathematically.

Cross-hemispheric integration plays a central role in these processes. While the left hemisphere is typically associated with sequential reasoning and linguistic organisation, and the right hemisphere with spatial awareness and holistic perception, creativity often arises when these complementary capacities interact recursively. In artistic composition or literary production, structural organisation may be guided by analytical processes while thematic development draws upon perceptual and emotional pattern recognition.

This synergistic interaction allows for the creation of works that resonate on multiple levels. Authors such as Salman Rushdie, for instance, maintain consistent narrative tone and thematic coherence across extensive texts by recursively engaging with linguistic and emotional patterns. Through repeated refinement, these patterns may be extended without losing continuity.

Recursive mirroring also contributes to the development of abstract thinking. Concepts such as success, failure, or love are not fixed representations but evolving configurations shaped through repeated engagement with sensory experience, emotional association, and contextual interpretation. Each reflective engagement introduces new layers of meaning, allowing abstract ideas to develop progressively greater depth.

Emotional patterns further enrich this process. Emotional responses may be regarded as patterned neural states that interact with cognitive representations. When engaging with artistic or literary works, individuals mirror the emotional configurations conveyed through the medium against their own experiential patterns. This recursive interaction deepens interpretative understanding and enhances engagement.

For example, when listening to a piece of contemporary music such as Lady Gaga's "Bad Romance", listeners may recognise familiar rhythmic or melodic elements derived from prior musical exposure. Simultaneously, the emotional intensity of the composition is mirrored against personal experience. With each recursive engagement, new layers of meaning and affective nuance may become apparent.

In the broader context of intellectual synergy, such emotional engagement facilitates the integration of analytical and intuitive processes. Creativity thus involves not only the generation of novel ideas but also the capacity to engage with them at both cognitive and affective levels.

As demonstrated throughout this chapter, creativity and intellectual synergy arise from the brain's recursive engagement with patterns across domains of experience. Through recognition, variation, and integration, these processes support the emergence of novel insights while preserving coherence with prior knowledge.

Subsequent chapters will explore how similar recursive mechanisms contribute to memory, learning, and social behaviour, as well as to the patterned organisation observed within biological systems. Understanding the relationship between creativity, intellectual synergy, and recursive mirroring provides insight into the processes through which human thought and cultural expression develop over time.

References

Boden, M.A., 2004. *The Creative Mind: Myths and Mechanisms*. 2nd ed. London: Routledge.

Csikszentmihalyi, M., 1996. *Creativity: Flow and the Psychology of Discovery and Invention*. New York: HarperCollins.

Dietrich, A., 2004. The cognitive neuroscience of creativity. *Psychonomic Bulletin and Review*, 11(6), pp.1011–1026.

Finke, R.A., Ward, T.B. and Smith, S.M., 1992. *Creative Cognition: Theory, Research, and Applications*. Cambridge, MA: MIT Press.

Gabora, L., 2010. Revenge of the 'neurds': Characterising creative thought in terms of the structure and dynamics of memory. *Creativity Research Journal*, 22(1), pp.1–13.

Guilford, J.P., 1950. Creativity. *American Psychologist*, 5(9), pp.444–454.

Kaufman, J.C. and Sternberg, R.J., 2010. *The Cambridge Handbook of Creativity*. Cambridge: Cambridge University Press.

Koestler, A., 1964. *The Act of Creation*. London: Hutchinson.

Mumford, M.D., Medeiros, K.E. and Partlow, P.J., 2012. Creative thinking: Processes, strategies, and knowledge. *Journal of Creative Behavior*, 46(1), pp.30–47.

Runco, M.A., 2007. *Creativity: Theories and Themes*. Amsterdam: Elsevier Academic Press.

Sawyer, R.K., 2011. *Explaining Creativity: The Science of Human Innovation*. 2nd ed. Oxford: Oxford University Press.

Simonton, D.K., 2012. Taking the U.S. Patent Office criteria seriously: A quantitative three-criterion creativity definition and its implications. *Creativity Research Journal*, 24(2–3), pp.97–106.

Sternberg, R.J. and Lubart, T.I., 1999. The concept of creativity: Prospects and paradigms. In: R.J. Sternberg, ed. *Handbook of Creativity*. Cambridge: Cambridge University Press, pp.3–15.

Ward, T.B., Smith, S.M. and Finke, R.A., 1999. *Creative Cognition*. Cambridge, MA: MIT Press.

Weisberg, R.W., 2006. *Creativity: Understanding Innovation in Problem Solving, Science, Invention, and the Arts*. Hoboken, NJ: Wiley.

Chapter 5:

Emotions, Memory, and Pattern Integration

Human experience is shaped by the interaction of emotions, memories, and cognitive processes, all of which are underpinned by the brain's ability to form, recognise, and refine patterns. While we often think of emotions as something separate from rational thought, the reality is that emotions are deeply intertwined with cognitive processes. They play a critical role in how we remember experiences, make decisions, and interact with the world around us. This chapter explores the role of emotional patterns in memory formation, the integration of these patterns with cognitive processes, and how recursive mirroring helps us navigate the complex landscape of human experience.

At the core of this discussion is the idea that emotions are not isolated feelings but patterned responses that the brain generates in reaction to external stimuli and internal states. Emotions are shaped by recurring patterns of experience, memory, and expectation, and they serve as powerful cues that guide behaviour and decision-making. The recursive nature of emotional processing allows the brain to mirror and refine emotional patterns over time, integrating them with cognitive patterns to form a cohesive understanding of both the self and the world.

One of the key functions of emotions is to enhance the brain's ability to recognise and store patterns in memory. When we experience strong emotions, the brain prioritises the encoding of those experiences, making it more likely that we will remember emotionally charged events. This is why memories associated with intense emotions, whether joy, fear, anger, or sadness, are often more vivid and lasting than neutral experiences. The emotional content of an experience acts as a marker, signalling to the brain that this memory is important and should be stored for future reference.

For instance, think back to a significant event in your life, such as a major accomplishment or a personal loss. What makes that memory stand out is not just the factual details but the emotions you felt at the time. The brain, through recursive mirroring, reflects these emotional patterns against cognitive patterns, creating a rich, layered memory that integrates both emotional and rational elements. This recursive integration allows the brain to encode memories in a way that makes them not only easier to recall but also more meaningful.

Memory, then, is not a passive recording of facts but an active, recursive process in which the brain mirrors and refines patterns of experience. Each time we recall a memory, the brain engages in a recursive process where it revisits the original pattern, often introducing slight variations based on new experiences or current emotional states. This recursive updating of memories is what allows us to integrate new information into our understanding of the past, making memory a dynamic and flexible system.

Emotional patterns, in this context, serve as anchors that shape how we recall and interpret memories. When we think of a past event, we do not simply remember what happened; we also re-experience the emotions associated with it. This recursive emotional mirroring enriches our understanding of the memory and allows it to take on new meaning over time. For example, a memory of a childhood fear may, in adulthood, be reframed through recursive emotional reflection, where the fear is tempered by new experiences of security and knowledge. The brain mirrors the original emotional pattern of fear, but with each reflection, it introduces new emotional layers, transforming the memory into something more nuanced.

This recursive integration of emotion and memory is also evident in how we process abstract concepts such as success or failure. These concepts are not simply cognitive ideas but are deeply connected to emotional patterns formed through past experiences. When we think about success, for instance, the brain mirrors not only the cognitive understanding of achievement but also the emotional patterns associated with moments of personal accomplishment, satisfaction, or pride. Each time we reflect on the concept of success, the brain engages in recursive mirroring, layering new emotional and cognitive patterns onto the concept, thereby refining and deepening our understanding of it.

Emotions, then, act as guiding patterns that help the brain make sense of complex, abstract ideas. This is particularly important in decision-making, where emotions often play a central role in guiding choices. The notion that human beings are purely rational decision-makers has long been challenged by the understanding that emotional patterns heavily influence our decisions. When faced with a choice, the brain mirrors past experiences and emotions, recursively refining the patterns until it arrives at a decision that feels right. This process is not always conscious but is often driven by gut feelings or intuition, both of which are shaped by the recursive interplay of emotional and cognitive patterns.

Take the example of someone deciding whether to accept a new job offer. On the surface, the decision may seem to involve logical considerations such as salary, career growth, and work-life balance. However, the emotional patterns associated with past experiences, perhaps a previous job that was stressful or a desire for greater fulfilment, will also play a significant role. The brain mirrors these emotional patterns against the current situation, introducing variations and dislocations as it weighs the emotional significance of the decision. This recursive process of integrating emotion with logic allows the individual to arrive at a decision that feels both rational and emotionally satisfying.

In this way, emotional processing and cognitive reasoning are not separate but are deeply interconnected through recursive mirroring. Emotions provide the motivational force behind many of our decisions and behaviours, while cognitive patterns offer the structure needed to analyse and evaluate choices. The brain's ability to mirror and integrate these patterns ensures that our behaviour is adaptive and appropriate to the context, whether we are making decisions in personal relationships, at work, or in social settings.

Another critical aspect of the recursive mirroring model is how the brain uses emotional patterns to anticipate and predict future outcomes. Just as we saw in previous chapters with pattern recognition and predictive coding, emotions play a key role in helping the brain anticipate how future events will unfold. When we face a potentially threatening situation, for example, the brain mirrors emotional patterns from past experiences of fear or danger and uses these patterns to predict the likelihood of harm. This recursive emotional reflection allows the brain to prepare an appropriate response, whether that means fighting, fleeing, or freezing.

This process is evident in social situations, where emotions serve as predictive cues that guide behaviour. In a conversation, for instance, we are not only processing the content of the other person's words but also mirroring their emotional cues, such as tone of voice, facial expressions, and body language, against our own emotional patterns. This recursive mirroring allows us to anticipate how the conversation will unfold and adjust our responses accordingly. Emotional patterns, in this context, act as a kind of social compass, helping us navigate complex interpersonal dynamics.

The brain's ability to integrate emotional and cognitive patterns is also what allows us to experience empathy. When we see someone in distress, our brain mirrors their emotional state, allowing us to feel what they are feeling. This empathetic mirroring is a recursive process where the brain reflects the emotional patterns of another person against our own emotional experiences, creating a deep connection between self and other. Empathy, in this sense, is a form of emotional pattern recognition that is essential for social cohesion and interpersonal relationships.

Memory also plays a vital role in cultural transmission. The brain's recursive ability to mirror and store emotional and cognitive patterns allows us to pass down knowledge, values, and traditions from one generation to the next. When we engage with cultural stories, myths, or rituals, we are not just learning new information but are also engaging with the emotional patterns embedded in these cultural practices. This recursive engagement with cultural patterns helps reinforce a sense of identity and belonging, ensuring that cultural knowledge is preserved and adapted over time.

In this way, the recursive mirroring model helps explain not only individual memory and emotion but also the transmission of emotional and cognitive patterns across generations. Whether we are reflecting on personal memories or engaging with cultural traditions, the brain's ability to integrate emotional and cognitive patterns ensures that our experiences are meaningful and enduring.

As we move into the next chapter, we will explore the biological foundations of these processes, examining how the brain's recursive mirroring of patterns reflects similar processes in nature. We will see how fractal patterns in biology and nature mirror the brain's ability to recognise and refine patterns, providing a deeper understanding of the recursive processes that underpin both biological and cognitive systems.

References

Barrett, L.F., 2017. *How Emotions Are Made: The Secret Life of the Brain*. London: Macmillan.

Damasio, A., 1994. *Descartes' Error: Emotion, Reason and the Human Brain*. London: Vintage.

Dolan, R.J., 2002. Emotion, cognition, and behaviour. *Science*, 298(5596), pp.1191–1194.

Dudai, Y., 2012. The restless engram: Consolidations never end. *Annual Review of Neuroscience*, 35, pp.227–247.

Eichenbaum, H., 2017. *Memory: Organization and Control*. 2nd ed. Oxford: Oxford University Press.

Frijda, N.H., 1986. *The Emotions*. Cambridge: Cambridge University Press.

Kensinger, E.A., 2009. Remembering the details: Effects of emotion. *Emotion Review*, 1(2), pp.99–113.

LeDoux, J., 1996. *The Emotional Brain*. New York: Simon and Schuster.

McGaugh, J.L., 2004. The amygdala modulates the consolidation of memories of emotionally arousing experiences. *Annual Review of Neuroscience*, 27, pp.1–28.

Nader, K. and Hardt, O., 2009. A single standard for memory: The case for reconsolidation. *Nature Reviews Neuroscience*, 10(3), pp.224–234.

Phelps, E.A., 2004. Human emotion and memory: Interactions of the amygdala and hippocampal complex. *Current Opinion in Neurobiology*, 14(2), pp.198–202.

Pessoa, L., 2008. On the relationship between emotion and cognition. *Nature Reviews Neuroscience*, 9(2), pp.148–158.

Rolls, E.T., 2014. *Emotion and Decision-Making Explained*. Oxford: Oxford University Press.

Schacter, D.L., 2012. *Searching for Memory: The Brain, the Mind, and the Past*. New York: Basic Books.

Tulving, E., 2002. Episodic memory: From mind to brain. *Annual Review of Psychology*, 53, pp.1–25.

Chapter 6:

Biological Analogies in Pattern Formation

The processes of recursive mirroring and pattern formation, central to the brain's cognitive and emotional functions, do not exist in isolation. They are mirrored in a broad range of biological systems, where similar recursive and patterned behaviours can be observed in nature. By examining biological analogies, we can deepen our understanding of how the brain mirrors complex patterns, adapts to new environments, and generates novel insights through recursive processes. In this chapter, we will explore the similarities between the brain's recursive pattern formation and the structures seen in fractal geometry, cellular behaviour, and organ regeneration. These analogies provide a framework for understanding how the recursive principles governing the brain's operations reflect larger, universal patterns found in biological and environmental systems.

Fractal patterns, often associated with the self-replicating, recursive structures found in nature, provide an excellent analogy for understanding how the brain processes and refines patterns. Fractals are shapes or structures that exhibit self-similarity at various scales, meaning that the pattern repeats itself regardless of the size at which it is observed. This is seen in the branching of trees, blood vessels, coastlines, and even in the neural networks of the brain. The recursive nature of fractals mirrors the brain's ability to repeat, mirror, and refine patterns across different levels of cognition and perception.

One of the most powerful examples of fractal geometry in nature is the branching pattern seen in trees. The fractal pattern of a tree allows it to efficiently distribute nutrients, absorb sunlight, and adapt to its environment. Each branch, from the largest to the smallest, mirrors the structure of the whole tree, exhibiting a recursive pattern that is repeated at every level. Similarly, in the brain, neural networks form fractal-like structures, with larger networks breaking down into smaller, self-similar units. This fractal organisation allows the brain to process complex information across multiple levels, from high-level abstract thought to basic sensory perception.

The branching pattern seen in blood vessels is another example of how fractal structures enhance biological efficiency. The recursive branching of veins and arteries ensures that blood is delivered to every part of the body in an optimised manner, just as the brain's recursive networks ensure the efficient processing of sensory and cognitive information. In both cases, the recursive nature of these systems allows for flexibility and adaptability, enabling the organism to respond to environmental challenges.

The fractal organisation of biological systems extends beyond physical structures to the behaviour of cells and organisms. One of the most striking examples of recursive pattern formation in biology is seen in the regenerative abilities of certain species, such as planarians and salamanders. These organisms possess the remarkable ability to regenerate entire limbs or organs, a process that mirrors the brain's capacity to form and refine patterns through recursive mirroring.

In the case of planarians, a small flatworm capable of regenerating its entire body from a single piece, the process of regeneration is driven by stem cells that mirror the original pattern of the organism. When a planarian is cut in half, the stem cells in each part begin to divide and differentiate, following the patterned blueprint of the whole organism. This recursive process of regeneration ensures that the new tissue mirrors the original structure, even as it introduces slight variations to adapt to the organism's current state.

Similarly, salamanders can regenerate lost limbs through a process that involves the formation of a blastema, a mass of undifferentiated cells that mirrors the structure of the original limb. The cells in the blastema divide and differentiate in a recursive manner, following the pattern of the limb while allowing for adjustments and refinements. This process is not unlike the brain's ability to mirror and refine cognitive patterns, introducing variations that allow for flexibility and adaptation.

The ability of certain organisms to regenerate limbs or organs reflects a broader principle of biological recursion, where patterns are mirrored, refined, and repeated across different levels of organisation.

Just as the brain mirrors and refines cognitive and emotional patterns through recursive processes, biological systems mirror and refine physical patterns to adapt to new challenges and ensure survival.

The brain's recursive processes, then, are not unique to cognition but are part of a larger, universal principle found throughout biology. This principle is rooted in the idea that adaptation and survival depend on the ability to recognise and mirror patterns, whether at the cellular level, the structural level, or the cognitive level. In this sense, the brain's recursive mirroring of patterns can be seen as a natural extension of the biological imperative to adapt, evolve, and refine in response to a changing environment.

The concept of symmetry is also critical to understanding how the brain mirrors biological patterns. Symmetry, in biological systems, refers to the balanced, proportional organisation of structures that allows for efficient functioning. This is evident in the bilateral symmetry of the human body, where the left and right sides mirror each other, creating a balanced structure that optimises movement, sensory processing, and survival. This symmetry is mirrored in the brain's structure, where the left and right hemispheres reflect and refine information in a recursive manner, allowing for complex cognitive processes.

Symmetry in biological systems is not just about physical balance; it also plays a role in development and evolution. In many species, the symmetry of an organism's body is a key factor in its ability to survive and reproduce. For example, in birds, the symmetry of wing structure is crucial for flight, ensuring that the animal can maintain balance and control in the air. In humans, the symmetry of facial features is often associated with attractiveness, as it signals genetic fitness and health. This emphasis on symmetry in biological systems mirrors the brain's own reliance on symmetrical, recursive processes to generate coherent thought, behaviour, and perception.

The recursive nature of symmetry in biological systems extends to evolutionary processes as well. Evolution, like the brain's cognitive processes, is driven by recursive mechanisms that introduce slight dislocations or variations in established patterns. These dislocations, much like mutations in DNA, create opportunities for adaptation and innovation. Over time, the recursive mirroring of patterns, combined with these dislocations, leads to the development of new traits, behaviours, and species.

In the same way, the brain's recursive mirroring of cognitive patterns allows for creativity and problem-solving by introducing slight variations that generate new insights and ideas. These variations, or dislocations, are not random but are part of a broader process of refinement and adaptation, ensuring that the brain can respond to novel challenges with flexibility and innovation.

Understanding the recursive nature of biological systems provides valuable insights into how the brain mirrors and refines patterns to create thought, behaviour, and perception. The fractal organisation of structures, the regenerative abilities of certain organisms, and the symmetry seen in both biology and cognition all point to a deeper, universal principle of pattern formation and recursion that governs both the physical and cognitive worlds.

As we move forward in this book, we will continue to explore how these biological analogies provide a framework for understanding the brain's operations. By examining the parallels between the brain's recursive processes and the structures found in nature, we gain a deeper appreciation of the fundamental patterns that underlie both biological life and human cognition.

In the next chapter, we will examine how these recursive processes extend to stereotyping and social behaviour, exploring how the brain's reliance on pattern recognition shapes the way we navigate the social world. We will also look at the evolutionary origins of these processes and their role in ensuring survival and adaptation in complex social environments.

References

Ball, P., 2009. *Shapes: Nature's Patterns*. Oxford: Oxford University Press.

Camazine, S., Deneubourg, J.L., Franks, N.R., Sneyd, J., Theraulaz, G. and Bonabeau, E., 2003. *Self-Organization in Biological Systems*. Princeton: Princeton University Press.

Forgacs, G. and Newman, S.A., 2005. *Biological Physics of the Developing Embryo*. Cambridge: Cambridge University Press.

Goodwin, B., 1994. *How the Leopard Changed Its Spots: The Evolution of Complexity*. London: Phoenix.

Kauffman, S.A., 1993. *The Origins of Order: Self-Organization and Selection in Evolution*. Oxford: Oxford University Press.

Levin, M., 2012. Morphogenetic fields in embryogenesis, regeneration, and cancer: Non-local control of complex patterning. *BioSystems*, 109(3), pp.243–261.

Newman, S.A. and Bhat, R., 2008. Dynamical patterning modules: A "pattern language" for development and evolution of multicellular form. *International Journal of Developmental Biology*, 52(5–6), pp.693–705.

Nijhout, H.F., 2001. *Elements of Developmental Biology*. New York: Springer.

Saló, E. and Baguñà, J., 2002. Regeneration in planarians and other worms: New findings, new tools, and new perspectives. *Journal of Experimental Zoology*, 292(6), pp.528–539.

Turing, A.M., 1952. The chemical basis of morphogenesis. *Philosophical Transactions of the Royal Society B*, 237(641), pp.37–72.

Werner, S. and Grose, R., 2003. Regulation of wound healing by growth factors and cytokines. *Physiological Reviews*, 83(3), pp.835–870.

Wolpert, L., 2011. *Principles of Development*. 4th ed. Oxford: Oxford University Press.

Chapter 7:

The Brain's Role in Stereotyping and Survival

The brain's ability to recognise and mirror patterns is not only essential for cognition and creativity but also plays a significant role in shaping social behaviour. One of the most powerful examples of this pattern recognition is the tendency for humans to engage in stereotyping, a process where the brain generalises information about a group or category based on limited experiences. While stereotyping is often viewed negatively in modern discourse, it is rooted in the brain's evolutionary need to quickly process and categorise information to ensure survival. This chapter explores the origins of stereotyping, how the brain's recursive processes contribute to this behaviour, and the role of stereotyping in human adaptation and social navigation.

At its core, stereotyping is a cognitive shortcut that allows the brain to make quick decisions by recognising and applying patterns. In evolutionary terms, this ability was crucial for survival. Early humans, for example, had to quickly assess whether a stranger or an unfamiliar group posed a threat. By relying on visual cues, behavioural patterns, and past experiences, the brain could make a generalised judgement, even with limited information. These generalisations, while not always accurate, allowed individuals to act swiftly in high-risk situations, minimising the likelihood of harm.

The brain's reliance on pattern recognition for stereotyping is grounded in the same recursive mirroring processes that govern other aspects of cognition. When the brain encounters a new person or situation, it mirrors the sensory input, such as facial features, clothing, or behaviour, against stored patterns of previous experiences. This mirroring allows the brain to make predictions based on generalised patterns associated with certain groups, behaviours, or environments. For example, if an individual has had multiple encounters with people from a particular group who have behaved aggressively, the brain may generalise this pattern to future encounters with people from the same group, anticipating similar behaviour.

This process of generalisation is not inherently negative but is a natural part of the brain's adaptive mechanisms. By forming generalised patterns, the brain conserves cognitive resources, allowing for quick and efficient decision-making in complex environments. The ability to stereotype, then, can be seen as part of the brain's broader strategy for navigating social interactions and ensuring survival in a world full of uncertainties. However, while stereotyping may have offered evolutionary advantages in early human societies, it also comes with significant limitations and can lead to bias and prejudice when applied rigidly.

One of the key reasons stereotyping can be problematic in modern society is that it often oversimplifies the complex nature of individual behaviour and identity. The brain's tendency to mirror and generalise patterns across groups can lead to a failure to recognise the individual nuances that exist within any category. When the brain mirrors a stereotype, whether based on race, gender, or socioeconomic status, it does so based on a limited set of experiences or societal norms, which may not accurately reflect the diversity within that group. This results in cognitive dislocation, where the stereotype no longer aligns with the actual behaviour or characteristics of individuals within the group.

The brain's recursive processes play a crucial role in both the formation and refinement of stereotypes. Initially, the brain may form a stereotype based on a small number of encounters or experiences, reflecting and mirroring those patterns in future interactions. However, with each new encounter, the brain introduces dislocations, slight variations in the pattern, allowing for the possibility of refinement. In this way, the brain's recursive mirroring of stereotypes is not fixed but dynamic, constantly being updated and adjusted based on new information.

This flexibility is what allows individuals to challenge and redefine stereotypes over time. When the brain encounters an individual who does not fit the established pattern of a stereotype, it mirrors this new information back against the old pattern, creating a dislocation that can lead to the modification of the stereotype. For example, if someone holds a stereotype about a particular ethnic group but then has a positive, meaningful interaction with a member of that group, the brain's recursive mirroring of this new experience can introduce a dislocation in the original stereotype, prompting a shift in perception.

In social contexts, stereotyping serves as a form of predictive coding, where the brain anticipates how others will behave based on generalised patterns. This predictive ability is essential for navigating complex social environments, especially when there is limited time or information available. For example, in a crowded marketplace or busy street, individuals may use stereotypes to quickly assess the intentions or behaviours of those around them, enabling them to navigate the situation more efficiently.

One of the most fascinating examples of this phenomenon can be observed in unregulated traffic environments, where drivers instinctively navigate chaotic situations without formal rules or traffic signals. In certain regions of the world, particularly in parts of Africa, India, and the Middle East, drivers rely on social cues and stereotypes about driving behaviour to avoid accidents and maintain the flow of traffic. Despite the absence of traffic lights, lane markings, or enforced regulations, drivers manage to pass through intersections, make turns, and negotiate with other drivers in a seemingly chaotic but surprisingly efficient manner.

In these environments, the brain relies heavily on pattern recognition and stereotyping to anticipate how other drivers will behave. By mirroring the behaviours of others, drivers develop an intuitive sense of the general rules governing the flow of traffic. For example, if drivers in a particular region tend to be more assertive at intersections, individuals may adjust their driving behaviour accordingly, relying on the stereotype of assertiveness to predict how others will navigate the situation. This form of instinctive cooperation emerges from the brain's ability to generalise and mirror patterns of behaviour, allowing for the spontaneous organisation of traffic flow without the need for formal regulation.

The success of these unregulated driving environments demonstrates how the brain's reliance on stereotyping and pattern recognition can lead to self-organising systems in which individuals instinctively cooperate to avoid collisions and maintain order. However, it also highlights the limitations of stereotyping, as this reliance on generalisations can sometimes lead to miscommunication or misjudgements, particularly when individual behaviours deviate from the established pattern.

While stereotyping plays an important role in social navigation, it is also closely linked to the brain's evolutionary need for survival. In the early days of human evolution, stereotyping was a protective mechanism that allowed individuals to assess potential threats quickly and efficiently. For example, early humans living in hunter-gatherer societies may have relied on visual cues, such as clothing, body language, or group affiliation, to determine whether another individual was a friend or foe. By generalising these patterns based on past experiences, early humans could make rapid decisions about whether to approach or avoid strangers, ensuring their survival in a dangerous and unpredictable environment.

This evolutionary legacy of stereotyping persists in modern society, even though the threats we face today are far more complex and nuanced. The brain's reliance on predictive coding and pattern generalisation continues to shape how we navigate social interactions, from everyday decisions about whom to trust to larger societal patterns involving race, gender, and class. While these generalisations can provide shortcuts in social decision-making, they also create challenges in a diverse and interconnected world where rigid stereotypes often fail to capture the richness and variability of human experience.

The brain's ability to modify stereotypes through recursive dislocations offers hope for overcoming the limitations of rigid generalisations. As individuals gain more exposure to diverse experiences and perspectives, the brain mirrors these new patterns, introducing variations that challenge and refine existing stereotypes. This recursive process is crucial for social learning and adaptation, as it allows individuals to move beyond simplistic generalisations and develop a more nuanced understanding of the world.

In the context of cultural evolution, the brain's reliance on stereotyping and pattern recognition can also be seen as part of a broader strategy for ensuring social cohesion and cooperation. Human societies have long relied on shared cultural patterns and norms to guide behaviour and maintain order. These patterns, much like stereotypes, are generalisations that provide a framework for understanding social roles, expectations, and interactions. However, just as the brain can modify individual stereotypes through recursive dislocations, societies can evolve their cultural norms and values over time, adapting to new challenges and embracing greater diversity.

In the following chapter, we will explore how the brain's recursive processes influence memory and learning, focusing on the role of neuroplasticity and the brain's ability to adapt and refine patterns through experience. We will examine how these processes shape not only individual learning but also the broader societal transmission of knowledge and culture.

References

Allport, G.W., 1954. *The Nature of Prejudice.* Cambridge, MA: Addison-Wesley.

Bargh, J.A., 1994. The four horsemen of automaticity: Awareness, intention, efficiency, and control in social cognition. In: R.S. Wyer and T.K. Srull, eds. *Handbook of Social Cognition.* Hillsdale, NJ: Lawrence Erlbaum Associates, pp.1–40.

Fiske, S.T. and Taylor, S.E., 2013. *Social Cognition: From Brains to Culture*. 2nd ed. London: Sage Publications.

Gigerenzer, G., 2007. *Gut Feelings: The Intelligence of the Unconscious*. London: Penguin Books.

Hamilton, D.L. and Sherman, S.J., 1996. Perceiving persons and groups. *Psychological Review*, 103(2), pp.336–355.

Haselton, M.G. and Nettle, D., 2006. The paranoid optimist: An integrative evolutionary model of cognitive biases. *Personality and Social Psychology Review*, 10(1), pp.47–66.

Kahneman, D., 2011. *Thinking, Fast and Slow*. London: Penguin Books.

Macrae, C.N. and Bodenhausen, G.V., 2000. Social cognition: Thinking categorically about others. *Annual Review of Psychology*, 51, pp.93–120.

Neuberg, S.L. and Cottrell, C.A., 2008. Managing the threats we pose to others: Threat management in social cognition. *Social and Personality Psychology Compass*, 2(3), pp.1315–1333.

Sherman, J.W., 2006. Automatic and controlled components of prejudice and stereotyping. *Psychological Inquiry*, 17(4), pp.299–314.

Tajfel, H., 1982. *Social Identity and Intergroup Relations*. Cambridge: Cambridge University Press.

Tversky, A. and Kahneman, D., 1974. Judgment under uncertainty: Heuristics and biases. *Science*, 185(4157), pp.1124–1131.

Zebrowitz, L.A., 1996. Physical appearance as a basis of stereotyping. In: C.N. Macrae, C. Stangor and M. Hewstone, eds. *Stereotypes and Stereotyping*. New York: Guilford Press, pp.79–120.

Chapter 8:

Memory, Learning, and Neuroplasticity

The brain's ability to form, refine, and adapt patterns is not only crucial for social navigation but also serves as the foundation for memory, learning, and adaptation. These processes are deeply intertwined with the brain's neuroplasticity, its capacity to reorganise and create new neural connections in response to experience and learning. In this chapter, we will explore how recursive pattern formation underpins memory and learning, how neuroplasticity enables the brain to constantly adapt, and how these processes shape both individual development and broader societal learning.

Memory, at its core, is the brain's ability to recognise, store, and retrieve patterns over time. It allows us to recall experiences, access knowledge, and draw connections between past and present. Memory is not a passive process where information is simply stored and later retrieved; rather, it is an active, recursive process where patterns are mirrored, refined, and even altered each time they are recalled. This continuous reworking of memories is a testament to the brain's dynamic nature, reflecting its ongoing adaptation and learning.

One of the key features of memory is its ability to integrate cognitive and emotional patterns. As we explored in earlier chapters, emotions serve as powerful anchors in the formation and recall of memories. A memory tied to a strong emotional experience, such as a significant personal achievement or a traumatic event, is often more vivid and persistent. This is because the brain prioritises the encoding of emotional experiences, mirroring these patterns against cognitive frameworks to create a lasting memory. Each time the memory is recalled, the brain mirrors the emotional pattern again, allowing for both reinforcement and refinement of the memory.

However, memory is also subject to dislocation, where the original pattern of a memory can shift or evolve based on new experiences or perspectives. This means that our memories are not static but are subject to constant revision. Each time we recall a memory, the brain introduces slight variations, mirroring the past against the present, and updating the memory to reflect new information or insights. This recursive process is why memories can change over time, why we may remember an event differently depending on how much time has passed or what new experiences have influenced our perspective.

This adaptive aspect of memory is closely linked to neuroplasticity, the brain's ability to reorganise its neural circuits in response to learning and experience. Neuroplasticity is the reason why the brain is able to learn new skills, recover from injury, and adapt to new environments. It reflects the brain's inherent flexibility, where existing patterns of neural activity can be refined, strengthened, or reorganised as needed.

Neuroplasticity allows the brain to create new synaptic connections between neurons, strengthening patterns that are frequently used while pruning those that are no longer relevant. This process ensures that the brain is constantly optimising its operations, focusing on the patterns that are most useful for navigating the world. For example, when we learn a new skill, such as playing a musical instrument or speaking a new language, the brain forms new neural circuits that mirror the patterns of the skill. Over time, these circuits become more refined as the brain recursively mirrors and practices the patterns, allowing for greater fluency and proficiency.

Learning, then, is not simply a matter of acquiring new information but involves the restructuring and refinement of existing patterns. The brain does not discard old patterns entirely but mirrors them against new information, creating recursive loops where knowledge is updated and reorganised. This is evident in conceptual learning, where the brain builds upon existing knowledge to form more complex and abstract ideas. Each new concept is mirrored against the old, introducing dislocations that refine and deepen understanding.

A powerful example of this process can be seen in mathematical learning. When students first encounter a mathematical concept, such as multiplication, their brains mirror this new pattern against the existing knowledge of addition. Over time, the brain refines its understanding of multiplication through recursive practice, integrating the pattern into a broader mathematical framework. As the student progresses to more complex mathematical concepts, such as algebra or calculus, the brain continues to mirror and refine these patterns, allowing for the gradual construction of a sophisticated mathematical understanding.

Neuroplasticity is also what enables the brain to recover from injury. In cases of traumatic brain injury or stroke, where specific neural pathways are damaged, the brain can often reorganise itself, forming new connections to compensate for the lost function. This process, known as functional plasticity, reflects the brain's remarkable ability to mirror and refine its patterns in response to challenges, ensuring that essential functions can be preserved or restored.

Learning is also a deeply social process, shaped not only by individual experiences but also by the transmission of knowledge and culture across generations. Just as the brain mirrors and refines patterns within itself, societies mirror and refine cultural patterns over time, ensuring the continuity and evolution of knowledge. This process is evident in the way cultural traditions, language, and social norms are passed down from one generation to the next. The brain's recursive processes, which allow for the integration of new information with existing patterns, are mirrored in the way societies adapt and refine their collective knowledge.

One of the most striking examples of social learning is the oral transmission of knowledge in cultures without written language. In these societies, knowledge is preserved through storytelling, songs, and rituals, which are repeated and mirrored across generations. Each retelling of a story introduces subtle variations, much like the brain's dislocation of patterns in memory. These variations reflect the evolving social context, allowing the story to remain relevant and meaningful while preserving its core message. This process of cultural recursion ensures that knowledge is both stable and adaptable, capable of evolving in response to new challenges or changing environments.

The brain's ability to learn and adapt is also influenced by environmental factors. Stimulating environments, where individuals are exposed to a variety of experiences, ideas, and challenges, promote greater neuroplasticity and enhance the brain's capacity to form and refine patterns. For example, children who grow up in environments rich in language, social interaction, and physical activity are more likely to develop robust neural networks that support cognitive flexibility and creativity. This is because the brain, when exposed to diverse patterns, mirrors these experiences and refines its neural circuits to optimise learning.

In contrast, deprived environments, where individuals have limited access to stimulation, can hinder neuroplasticity and reduce the brain's capacity to adapt. Research has shown that children raised in environments with limited sensory, social, or intellectual stimulation may develop weaker neural circuits, making it more difficult for them to learn new skills or adapt to new situations later in life. This highlights the importance of environmental enrichment in promoting brain development and lifelong learning.

The recursive nature of learning also extends to education systems, where teaching methods often mirror the brain's processes of pattern recognition and refinement. Effective teaching involves not just the transmission of information but the scaffolding of knowledge, where students are guided through increasingly complex patterns of understanding. This approach mirrors the brain's own recursive processes, where learning builds on existing patterns and is refined through practice and reflection.

For example, in a classroom setting, teachers often introduce new concepts by first connecting them to students' prior knowledge. This allows students to mirror the new information against familiar patterns, creating a foundation for deeper understanding. As students engage with the material through practice, discussion, and application, their brains recursively mirror and refine these patterns, leading to mastery of the subject. This process is not linear but involves constant feedback loops, where students test their understanding, receive feedback, and adjust their patterns of thinking accordingly.

The brain's recursive processes of learning and memory are also mirrored in the way societies organise knowledge. The development of libraries, universities, and research institutions reflects a broader societal effort to preserve, refine, and transmit patterns of knowledge across generations. Just as the brain mirrors and refines patterns within its neural circuits, societies mirror and refine knowledge through the creation of archives, educational curricula, and scientific research. This ensures that knowledge is not only preserved but also adapted to meet the needs of new generations and evolving challenges.

In summary, the brain's ability to form and refine patterns through neuroplasticity underpins both individual learning and the broader transmission of knowledge within societies. Memory, learning, and adaptation are dynamic processes, shaped by the brain's recursive mirroring of patterns and its capacity for constant reorganisation. Whether we are learning a new skill, recovering from an injury, or preserving cultural traditions, the brain's recursive processes ensure that knowledge is continuously refined, updated, and passed on.

In the next chapter, we will explore how these recursive processes manifest in language and literature, examining how the brain's ability to mirror and refine patterns shapes the way we communicate, create meaning, and understand complex narratives.

References

Doidge, N., 2007. *The Brain That Changes Itself.* London: Penguin Books.

Draganski, B. and May, A., 2008. Training-induced structural changes in the adult human brain. *Behavioural Brain Research*, 192(1), pp.137–142.

Dudai, Y., 2004. *The Neurobiology of Consolidations: Or, How Stable is the Engram?* Annual Review of Psychology, 55, pp.51–86.

Hebb, D.O., 1949. *The Organisation of Behaviour: A Neuropsychological Theory.* New York: Wiley.

Kandel, E.R., 2001. The molecular biology of memory storage: A dialogue between genes and synapses. *Science*, 294(5544), pp.1030–1038.

Kolb, B. and Gibb, R., 2011. Brain plasticity and behaviour in the developing brain. *Journal of the Canadian Academy of Child and Adolescent Psychiatry*, 20(4), pp.265–276.

Maguire, E.A., Woollett, K. and Spiers, H.J., 2006. London taxi drivers and bus drivers: A structural MRI and neuropsychological analysis. *Hippocampus*, 16(12), pp.1091–1101.

Merzenich, M.M., 2013. *Soft-Wired: How the New Science of Brain Plasticity Can Change Your Life.* San Francisco: Parnassus Publishing.

Nader, K., Schafe, G.E. and LeDoux, J.E., 2000. Fear memories require protein synthesis in the amygdala for reconsolidation after retrieval. *Nature*, 406(6797), pp.722–726.

Pascual-Leone, A., Amedi, A., Fregni, F. and Merabet, L.B., 2005. The plastic human brain cortex. *Annual Review of Neuroscience*, 28, pp.377–401.

Rosenzweig, M.R. and Bennett, E.L., 1996. Psychobiology of plasticity: Effects of training and experience on brain and behaviour. *Behavioural Brain Research*, 78(1), pp.57–65.

Squire, L.R. and Kandel, E.R., 2009. *Memory: From Mind to Molecules*. 2nd ed. New York: Roberts and Company.

Tulving, E. and Craik, F.I.M., 2000. *The Oxford Handbook of Memory*. Oxford: Oxford University Press.

Chapter 9:

The Role of Recursive Patterns in Language and Literature

Language is one of the most profound expressions of the brain's ability to mirror and refine patterns, and it serves as the foundation for communication, cultural transmission, and creative expression. Through recursive processes, the brain is able to organise sounds, symbols, and meanings into coherent linguistic structures, allowing us to convey thoughts, emotions, and abstract concepts. In this chapter, we will explore how the brain's pattern recognition and recursive mirroring processes underlie the structure and function of language. We will also examine how these processes shape literature and storytelling, giving rise to complex narratives that resonate with readers and communicate meaning across cultures and time.

At its core, language is a system of patterns, where sounds or symbols are arranged according to specific rules to create meaning. The brain's ability to understand and produce language relies on its capacity to recognise, mirror, and refine these linguistic patterns. From the simplest sentence structures to the most intricate poetic forms, language involves the recursive layering of sounds, words, and ideas. This recursive nature allows for the creation of infinite meanings from a finite set of symbols, a principle famously captured by Noam Chomsky's theory of generative grammar, which highlights how humans can generate an infinite number of sentences using a limited number of grammatical rules.

The brain's recursive processes begin with the recognition of phonetic patterns, the basic sounds that make up spoken language. When we hear someone speak, our brain immediately begins to mirror the phonetic patterns of their speech, comparing them to stored patterns of familiar sounds. This allows us to decode the sounds into words, a process that happens almost instantaneously. The ability to recognise and mirror these phonetic patterns is essential for understanding language, and it is a skill that develops early in childhood through exposure to speech.

As the brain mirrors phonetic patterns, it also begins to recognise morphological patterns, the building blocks of words. Morphology refers to the way words are structured and how their parts, such as prefixes, roots, and suffixes, combine to create meaning. The brain uses morphological mirroring to understand how words change in form depending on their function in a sentence. For instance, the brain recognises that adding "-ed" to a verb indicates that the action took place in the past. This recursive understanding of word formation allows us to interpret the meaning of sentences even when we encounter unfamiliar words, as we can often deduce their meaning from the morphological patterns they follow.

Beyond individual words, the brain engages in recursive pattern recognition at the level of syntax, the set of rules that governs how words are combined into sentences. Syntax operates through recursive structures, where clauses can be nested within other clauses, allowing for complex sentence formation. For example, in the sentence "The boy who won the race was happy," the phrase "who won the race" is a relative clause nested within the main clause. The brain mirrors these syntactic patterns as we construct and understand sentences, allowing us to comprehend complex ideas through recursive linguistic structures.

This recursive layering of phonetics, morphology, and syntax forms the basis of linguistic communication. However, language is more than just a system of rules; it is also a tool for conveying meaning, emotion, and intention. The brain's ability to mirror linguistic patterns allows it to integrate not only the structural aspects of language but also the semantic (meaning-based) and pragmatic (context-based) dimensions. This integration is what enables us to understand not only the literal meaning of words but also the implied meaning, tone, and subtext behind language.

For instance, when we hear the phrase "It's raining cats and dogs," the brain does not interpret this literally but instead mirrors the semantic pattern of the idiom, recognising that it is a metaphor for heavy rain. This ability to recognise and interpret metaphors, idioms, and other figurative language relies on the brain's recursive capacity to layer meanings and compare them to familiar patterns. Each time we encounter figurative language, the brain mirrors it against previous experiences of similar phrases, refining its understanding of the speaker's intention.

The recursive processes that govern language also play a central role in literature and storytelling, where patterns of meaning, emotion, and structure are mirrored and refined to create complex narratives. In literature, writers use linguistic patterns, such as rhyme, meter, alliteration, and metaphor, to engage the reader's brain in a recursive process of pattern recognition and interpretation. This recursive engagement allows readers to find deeper layers of meaning as they reflect on the text, mirroring the patterns of the story against their own experiences and emotions.

Consider the use of repetition in poetry, a technique where words, phrases, or sounds are repeated at regular intervals to create rhythm and emphasise key themes. Repetition acts as a recursive tool, allowing the reader's brain to mirror the repeated pattern and recognise its importance. For example, in T.S. Eliot's poem "The Love Song of J. Alfred Prufrock," the repeated line "In the room the women come and go / Talking of Michelangelo" creates a sense of rhythm and continuity, while also highlighting the protagonist's sense of social isolation and self-consciousness. Each time the line is repeated, it gains new significance as the reader mirrors it against the unfolding narrative.

Another powerful example of recursive patterning in literature is the use of parallelism and chiasmus, where ideas or structures are mirrored within the text. Parallelism involves repeating a similar grammatical structure, while chiasmus inverts the order of elements for rhetorical effect. These techniques engage the reader's brain in a process of reflection and comparison, allowing for the layering of meaning. For instance, Charles Dickens often used parallel structures to contrast social conditions, as seen in the famous opening lines of *A Tale of Two Cities*: "It was the best of times, it was the worst of times, it was the age of wisdom, it was the age of foolishness." The mirrored structure of these phrases allows the brain to compare and contrast the conditions, deepening the reader's understanding of the dualities present in the story.

Recursive mirroring in literature is also evident in the way narratives are structured, particularly in nonlinear storytelling or narratives that involve flashbacks, repetitions, and mirrored motifs. Authors like Salman Rushdie, who we discussed earlier, are masters of using recursive narrative techniques to create intricate layers of meaning. In his novel *Midnight's Children*, for example, Rushdie mirrors the history of India's independence with the personal history of the protagonist, Saleem Sinai. The story constantly shifts between historical events and personal memories, creating a recursive structure where the reader's understanding of the narrative is continuously refined with each new revelation.

This recursive interplay between memory, history, and personal experience in literature mirrors the brain's own processes of memory retrieval and pattern refinement. Just as the brain recursively mirrors past experiences against present ones, authors use recursive storytelling to create narratives that invite readers to reflect on how the past shapes the present and how patterns of behaviour, emotion, and thought are repeated across time.

The role of recursive patterns in literature is not limited to high art or complex narratives; it is also present in everyday storytelling and communication. When we tell stories to one another, we often rely on familiar patterns of narrative structure, beginning, middle, and end, that help listeners follow and understand the story. These narrative patterns are recursive, as they are mirrored across different cultures and time periods, allowing for the transmission of universal themes and lessons. The brain's ability to recognise and mirror these narrative patterns enables us to connect with stories on a deep emotional and cognitive level, whether we are reading a novel, listening to a friend's anecdote, or watching a film.

In oral storytelling traditions, recursive patterns are particularly important for ensuring the continuity and preservation of knowledge. In many indigenous cultures, knowledge is passed down through stories that are repeated across generations. These stories often contain repeated motifs, phrases, or structures that make them easier to remember and transmit. The brain mirrors these repeated patterns, allowing storytellers to recall and convey the narrative with remarkable accuracy. Each retelling introduces subtle variations, mirroring the brain's recursive dislocation of patterns in memory, ensuring that the story remains relevant to the current social or cultural context.

The recursive nature of language and literature is also evident in the way readers and listeners engage with stories. As we read or listen to a story, our brain mirrors the emotional patterns of the characters, allowing us to experience empathy and emotional resonance. This process of empathic mirroring is recursive, as it involves the continuous reflection of the characters' emotions against our own, deepening our connection to the narrative. The more we engage with the story, the more our brain refines these emotional patterns, allowing for a deeper understanding of the characters' motivations, struggles, and growth.

In summary, the brain's ability to recognise, mirror, and refine linguistic patterns is central to both language and literature. Through recursive processes, the brain is able to construct meaning from sounds, words, and sentences, as well as from complex narratives that engage us emotionally and intellectually. Literature, in particular, uses recursive structures to create layers of meaning that invite readers to reflect on the patterns of human experience, history, and emotion. Whether we are engaging with poetry, novels, or oral storytelling traditions, the brain's recursive mirroring of patterns allows us to connect with language and literature in profound and meaningful ways.

In the following chapter, we will explore how these recursive processes manifest in music and the synergy of patterns, examining how the brain's recognition of musical patterns influences emotional experience, creativity, and social cohesion.

References

Bruner, J., 1990. *Acts of Meaning*. Cambridge, MA: Harvard University Press.

Chomsky, N., 1965. *Aspects of the Theory of Syntax*. Cambridge, MA: MIT Press.

Clark, A., 1997. *Being There: Putting Brain, Body, and World Together Again*. Cambridge, MA: MIT Press.

Fauconnier, G. and Turner, M., 2002. *The Way We Think: Conceptual Blending and the Mind's Hidden Complexities*. New York: Basic Books.

Gentner, D. and Goldin-Meadow, S., 2003. *Language in Mind: Advances in the Study of Language and Thought*. Cambridge, MA: MIT Press.

Lakoff, G. and Johnson, M., 1980. *Metaphors We Live By*. Chicago: University of Chicago Press.

Pinker, S., 1994. *The Language Instinct*. New York: William Morrow.

Rumelhart, D.E., 1975. Notes on a schema for stories. In: D.G. Bobrow and A.M. Collins, eds. *Representation and Understanding*. New York: Academic Press, pp.211–236.

Schank, R.C. and Abelson, R.P., 1977. *Scripts, Plans, Goals and Understanding*. Hillsdale, NJ: Lawrence Erlbaum Associates.

Turner, M., 1996. *The Literary Mind*. Oxford: Oxford University Press.

Vygotsky, L.S., 1962. *Thought and Language*. Cambridge, MA: MIT Press.

Winograd, T., 1972. *Understanding Natural Language*. New York: Academic Press.

Zwaan, R.A. and Radvansky, G.A., 1998. Situation models in language comprehension and memory. *Psychological Bulletin*, 123(2), pp.162–185.

Chapter 10:

Music and the Synergy of Patterns

Music, much like language, is an extraordinary example of the brain's ability to recognise, mirror, and refine patterns. It taps into the deepest layers of human cognition and emotion, creating a synergy of patterns that resonate both intellectually and emotionally. The brain's response to music is driven by its capacity to detect rhythmic, harmonic, and melodic patterns, but what makes music truly powerful is the way these patterns interact and create complex emotional responses. In this chapter, we will explore how the brain recognises and processes musical patterns, how recursive mirroring contributes to musical creativity, and how music fosters social cohesion and emotional resonance through the synergy of patterns.

At the most fundamental level, music is composed of patterns of sound, arranged in ways that are meaningful to the human brain. These patterns include rhythms, melodies, and harmonies, all of which the brain is adept at recognising and mirroring. When we listen to a piece of music, the brain quickly begins to mirror the rhythmic patterns of the sounds, creating a sense of anticipation and predictive coding. This ability to anticipate the beat or the next note is central to the pleasure we derive from music, as it allows the brain to experience expectation and satisfaction when the pattern unfolds as predicted.

Rhythmic patterns, in particular, engage the brain's motor systems, creating a physical response to music. Even when we are sitting still, the brain's motor cortex is activated as it mirrors the rhythmic patterns of the music, creating an impulse to move. This is why we often feel the urge to tap our feet or nod our heads in time with the beat, even when we are not consciously aware of doing so. The brain's ability to mirror rhythmic patterns is not just a cognitive process but also a bodily one, connecting music to the physical experience of movement and coordination.

Beyond rhythm, the brain also mirrors melodic and harmonic patterns in music. Melody refers to the sequence of notes that form a recognisable tune, while harmony refers to the combination of different notes played simultaneously to create a rich, layered sound. The brain's ability to recognise and mirror these patterns allows us to appreciate the complexity of music, as it processes both the horizontal dimension (melody) and the vertical dimension (harmony) simultaneously. This multidimensional processing is what gives music its emotional depth, as the brain layers the patterns of melody and harmony together to create a synergistic experience.

One of the most fascinating aspects of music is its ability to evoke emotions. The brain's response to music is deeply emotional, and this is largely due to its recursive mirroring of musical patterns. Each time we listen to a piece of music, the brain mirrors the emotional patterns embedded in the sound, whether through changes in tempo, key, or dynamics, and reflects these patterns against our own emotional experiences. This recursive emotional mirroring creates a powerful connection between the listener and the music, allowing us to feel a wide range of emotions, from joy and excitement to melancholy and nostalgia.

For example, when we listen to a minor key melody, the brain mirrors the sombre and introspective qualities of the music, often evoking feelings of sadness or reflection. Conversely, a major key melody with an upbeat tempo might elicit feelings of happiness or energy. These emotional responses are not arbitrary but are rooted in the brain's ability to recognise and mirror the patterns of tension and resolution within the music. The brain anticipates the harmonic progression of a piece, creating tension when a dissonant chord is played and releasing that tension when the harmony resolves to a consonant chord. This recursive build-up and release of tension is what gives music its emotional power, as the brain mirrors the patterns of the music in real time.

The emotional synergy created by music is further enhanced by the brain's ability to layer patterns from different sensory modalities. When we attend a live concert, for instance, the brain does not only mirror the auditory patterns of the music but also integrates visual patterns from the performance, such as the movement of the musicians, the lighting, and the atmosphere of the venue. This multisensory integration creates a richer and more immersive experience, as the brain mirrors and layers the patterns from both the auditory and visual domains, creating a powerful emotional response that resonates long after the performance has ended.

Music also has a unique ability to evoke memories, which is another aspect of the brain's recursive patterning. When we hear a familiar song from our past, the brain mirrors the musical patterns against our stored memories of when we first heard the song. This recursive reflection allows the music to evoke not only the memory of the song but also the emotions and experiences associated with that time in our lives. This is why certain songs can instantly transport us back to specific moments, such as a childhood memory, a significant relationship, or a life-changing event. The brain's ability to layer musical patterns with emotional and autobiographical memories enhances the emotional resonance of music, making it a powerful tool for both personal reflection and emotional healing.

Musical creativity is another area where the brain's recursive mirroring processes play a crucial role. Composers and musicians often engage in a recursive process of mirroring familiar musical patterns while introducing slight dislocations or variations to create something new. This is similar to the pattern dislocation we discussed earlier in relation to creativity in the arts and sciences. By building on established musical forms, such as sonatas, symphonies, or jazz progressions, and then introducing novel variations, musicians are able to push the boundaries of musical expression while still maintaining a sense of coherence.

One striking example of this recursive creativity is found in the work of Beethoven, whose symphonies are known for their evolving, metamorphic structures. Beethoven would often introduce a musical motif early in a symphony, only to mirror and refine that motif through recursive variations as the piece progressed. This recursive development of themes allowed Beethoven to create music that was both recognisable and continually evolving, reflecting the brain's own processes of mirroring and refining patterns over time.

In more improvisational forms of music, such as jazz, musicians engage in a highly dynamic form of recursive mirroring. Jazz improvisation involves taking a familiar musical pattern, such as a chord progression or melodic line, and then introducing spontaneous variations, often in response to the patterns played by other musicians. This process of call and response, where musicians mirror and reflect each other's playing, creates a recursive loop that allows the music to evolve in real time. The brain's ability to mirror these patterns and respond with creative variations is what gives jazz its sense of spontaneity and freedom.

Music's role in social cohesion is another fascinating area where the brain's recursive mirroring processes come into play. Throughout history, music has been used as a tool for bringing people together, whether in religious rituals, social gatherings, or political movements. This is because music has the unique ability to create shared patterns of experience, allowing individuals to synchronise their emotions, movements, and behaviours with others. When people sing or dance together, the brain mirrors the rhythmic and melodic patterns of the music, creating a sense of collective unity and emotional bonding.

This synchronisation is not just metaphorical; it is also physical. When people listen to music together, their heart rates and breathing patterns often synchronise, reflecting the brain's ability to mirror and align with the rhythmic patterns of the music. This physical synchronisation enhances the sense of emotional connection and solidarity, as individuals become part of a larger collective experience. In this way, music fosters social cohesion by creating a shared emotional and physical experience, allowing people to feel connected to one another in ways that transcend language or cultural differences.

The brain's ability to create synergy through recursive mirroring of musical patterns also plays a role in therapeutic settings. Music therapy is a growing field that harnesses the emotional and cognitive power of music to help individuals with a wide range of conditions, from anxiety and depression to autism and Alzheimer's disease. By engaging patients in musical activities, such as listening, singing, or playing instruments, therapists can help individuals reconnect with emotional patterns, enhance memory, and improve social interaction.

The brain's recursive mirroring of musical patterns in these settings promotes neuroplasticity, helping patients adapt to challenges and recover lost cognitive or emotional functions.

In summary, music is a profound example of the brain's ability to create synergy through the recursive mirroring and refinement of patterns. Whether through rhythmic anticipation, harmonic resolution, or emotional resonance, music engages the brain's pattern recognition systems in ways that are both cognitive and emotional. It fosters creativity by allowing for the recursive dislocation of familiar patterns, and it promotes social cohesion by creating shared emotional and physical experiences. Music's power to evoke memories, generate emotions, and bring people together is rooted in the brain's recursive processes, making it a universal language that transcends cultural and temporal boundaries.

In the next chapter, we will continue exploring how these recursive processes manifest in art and architecture, focusing on how the brain recognises and responds to visual patterns and symmetry in the world around us.

References

Blood, A.J. and Zatorre, R.J., 2001. Intensely pleasurable responses to music correlate with activity in brain regions implicated in reward and emotion. *Proceedings of the National Academy of Sciences*, 98(20), pp.11818–11823.

Deutsch, D., 2013. *The Psychology of Music*. 3rd ed. London: Academic Press.

Huron, D., 2006. *Sweet Anticipation: Music and the Psychology of Expectation*. Cambridge, MA: MIT Press.

Juslin, P.N. and Västfjäll, D., 2008. Emotional responses to music: The need to consider underlying mechanisms. *Behavioral and Brain Sciences*, 31(5), pp.559–575.

Koelsch, S., 2014. Brain correlates of music-evoked emotions. *Nature Reviews Neuroscience*, 15(3), pp.170–180.

Levitin, D.J., 2006. *This Is Your Brain on Music*. London: Atlantic Books.

Lerdahl, F. and Jackendoff, R., 1983. *A Generative Theory of Tonal Music*. Cambridge, MA: MIT Press.

Patel, A.D., 2008. *Music, Language, and the Brain*. Oxford: Oxford University Press.

Peretz, I. and Zatorre, R.J., 2005. Brain organisation for music processing. *Annual Review of Psychology*, 56, pp.89–114.

Sacks, O., 2007. *Musicophilia: Tales of Music and the Brain*. London: Picador.

Sloboda, J.A., 1985. *The Musical Mind: The Cognitive Psychology of Music*. Oxford: Oxford University Press.

Zatorre, R.J., Chen, J.L. and Penhune, V.B., 2007. When the brain plays music: Auditory–motor interactions in music perception and production. *Nature Reviews Neuroscience*, 8(7), pp.547–558.

Chapter 11:

Art, Architecture, and the Brain's Visual Patterns

The human brain is profoundly attuned to recognising and processing visual patterns, whether in the natural world or in artistic and architectural forms. Visual patterns, such as symmetry, proportion, and fractal structures, evoke powerful cognitive and emotional responses, much like the patterns found in music. In this chapter, we will explore how the brain recognises and responds to visual patterns, how recursive processes are involved in the creation and appreciation of art and architecture, and how these visual structures reflect the brain's innate drive to find order, balance, and meaning in the world around us.

Visual art, at its most fundamental level, is a reflection of the brain's ability to mirror and refine patterns. When we look at a painting, sculpture, or architectural design, our brain immediately begins to recognise the shapes, lines, and colours that form the composition. This process of visual pattern recognition engages the brain's occipital lobe, responsible for processing visual stimuli, and the parietal and temporal lobes, which integrate this information with memory, emotion, and spatial reasoning. As the brain mirrors these visual patterns against its stored knowledge of past experiences and cultural associations, it generates emotional and intellectual responses that shape our perception of the artwork.

One of the most powerful visual patterns that the brain responds to is symmetry. Symmetry is found throughout nature, from the bilateral symmetry of the human body to the radial symmetry of flowers and the intricate structures of snowflakes. The brain is naturally drawn to symmetrical patterns because they signify order and balance, qualities that are essential for survival in both the natural and social world. Symmetry allows the brain to quickly and efficiently process information, as symmetrical objects and forms are easier to recognise and interpret than asymmetrical ones.

In art and architecture, symmetry has long been used to create visual harmony and evoke feelings of beauty and proportion. From the perfectly balanced proportions of classical Greek sculpture to the symmetrical floor plans of Renaissance cathedrals, artists and architects have used symmetry to create works that resonate deeply with the brain's desire for order and balance. When we look at a symmetrical work of art or building, the brain mirrors the visual patterns, reflecting them back and forth between the hemispheres and generating a sense of coherence and satisfaction.

Leonardo da Vinci famously studied symmetry and proportion in his artistic works, most notably in his drawing Vitruvian Man, which illustrates the ideal human proportions based on the symmetrical relationships of the body's parts. Da Vinci's use of symmetry in both his art and scientific observations reflects the brain's own drive to find patterned order in the world, mirroring the structures of nature and translating them into human creations.

While symmetry creates a sense of harmony, it is the brain's ability to recognise and mirror dislocated patterns, those that break or deviate from symmetry, that often creates intrigue and emotional impact in art. Asymmetry and dislocation introduce tension into a composition, capturing the viewer's attention by challenging the brain's expectations. This tension can evoke a range of emotional responses, from discomfort to fascination, as the brain mirrors the dislocated patterns and attempts to reconcile them with its internal sense of order.

For example, Pablo Picasso's work in Cubism represents a radical departure from traditional symmetrical forms, introducing fragmented and dislocated visual patterns that force the viewer to engage in recursive mirroring and reinterpretation. In a painting like *Les Demoiselles d'Avignon*, Picasso dislocates the human figure into angular shapes and fractured planes, challenging the brain's expectations of how the body should be represented. This dislocation creates a sense of visual tension, as the brain mirrors the disrupted patterns and attempts to make sense of them within its existing framework of human anatomy and artistic conventions.

Similarly, modern architecture, especially in the works of architects like Frank Gehry and Zaha Hadid, often plays with asymmetry and dislocation to create buildings that defy traditional expectations of form and function. Gehry's Guggenheim Museum Bilbao, with its flowing, curvilinear shapes and dislocated angles, forces the brain to mirror these unexpected patterns, creating a sense of movement and dynamism that challenges our conventional understanding of architecture. These dislocated patterns in modern architecture engage the brain in a recursive process of reflection and reinterpretation, much like abstract art, leading to new insights and emotional responses.

Beyond symmetry and dislocation, the brain also responds to fractal patterns in visual art and architecture. Fractals, as discussed in earlier chapters, are self-similar patterns that repeat at different scales, creating complex, recursive structures. In nature, fractal patterns are found in clouds, coastlines, trees, and mountains, and they reflect the underlying recursive processes that govern many natural phenomena. The brain's ability to recognise and mirror fractal patterns is linked to its capacity for recursive thinking, allowing us to see the connections between large-scale structures and their smaller components.

In art, fractal patterns are often used to create a sense of infinite complexity and depth. For example, the intricate designs of Islamic geometric art often feature repeating fractal patterns that mirror each other at different scales, creating a sense of both order and complexity. The brain mirrors these recursive patterns, reflecting them across its visual and cognitive domains, leading to a sense of awe and wonder at the intricate beauty of the design.

In architecture, fractal patterns are also employed to create structures that feel organic and natural. Traditional Indian temples often feature fractal designs, where the architectural elements repeat at different scales, from the smallest carvings to the overall structure of the building. This recursive use of fractals not only creates a visually harmonious design but also reflects the brain's desire to find patterns that extend across different levels of complexity.

The brain's response to proportion and scale is another important aspect of how we perceive art and architecture. The concept of proportion has been central to artistic and architectural design for centuries, with many cultures adhering to specific proportional systems, such as the Golden Ratio, to create visually pleasing compositions. The Golden Ratio, a mathematical proportion found in nature, has been used in everything from the design of ancient Greek temples to the composition of Renaissance paintings. The brain's recognition of these proportional patterns creates a sense of visual harmony, as the recursive relationship between the parts and the whole reflects the natural balance found in living systems.

In architecture, the use of proportion and scale can dramatically influence how we experience a space. Buildings designed with harmonious proportions, such as Notre-Dame Cathedral or the

Parthenon, evoke feelings of grandeur and balance, as the brain mirrors the proportional relationships between the architectural elements and the overall structure. Conversely, modern architectural designs that play with scale, such as the towering skyscrapers of New York or Dubai, create a sense of awe or even disorientation, as the brain attempts to reconcile the vast scale of the building with its human experience.

Light and shadow also play a crucial role in how the brain perceives visual patterns in art and architecture. The interplay of light and shadow can enhance or disrupt visual symmetry, create depth, and evoke emotional responses. In Baroque art, for example, artists like Caravaggio used dramatic contrasts of light and shadow, known as chiaroscuro, to heighten the emotional intensity of their scenes. The brain mirrors these visual contrasts, creating a sense of drama and tension as it interprets the light patterns within the composition.

In architecture, light is often used to shape how we experience a space. The play of light and shadow in buildings designed by architects like Louis Kahn or Tadao Ando creates dynamic spaces that change throughout the day, engaging the brain in a recursive process of visual reinterpretation. The shifting patterns of light create a living, evolving experience that mirrors the brain's capacity to recognise and adapt to changing visual stimuli.

In summary, the brain's response to art and architecture is deeply rooted in its ability to recognise, mirror, and refine visual patterns.

Whether through the symmetry of classical design, the dislocation of modern art, or the recursive complexity of fractal structures, the brain engages with visual patterns on multiple levels, creating emotional, intellectual, and spatial experiences that resonate deeply with human perception. Art and architecture, in turn, reflect the brain's own drive to find meaning, order, and beauty in the world through the recursive mirroring of visual patterns.

In the next chapter, we will explore how these recursive processes manifest in social behaviour, particularly in the ways that individuals and groups synchronise their actions, emotions, and responses through shared patterns of behaviour and communication.

Reference List

Arnheim, R., 1974. *Art and Visual Perception: A Psychology of the Creative Eye*. Berkeley: University of California Press.

Chatterjee, A., 2014. *The Aesthetic Brain: How We Evolved to Desire Beauty and Enjoy Art*. Oxford: Oxford University Press.

Gombrich, E.H., 1960. *Art and Illusion: A Study in the Psychology of Pictorial Representation*. London: Phaidon.

Gregory, R.L., 1997. *Eye and Brain: The Psychology of Seeing*. 5th ed. Oxford: Oxford University Press.

Hoffman, D.D., 1998. *Visual Intelligence: How We Create What We See*. New York: W.W. Norton.

Jacobsen, T., 2006. Bridging the arts and sciences: A framework for the psychology of aesthetics. *Leonardo*, 39(2), pp.155–162.

Koffka, K., 1935. *Principles of Gestalt Psychology*. London: Routledge.

Livingstone, M., 2002. *Vision and Art: The Biology of Seeing*. New York: Abrams.

Ramachandran, V.S. and Hirstein, W., 1999. The science of art: A neurological theory of aesthetic experience. *Journal of Consciousness Studies*, 6(6–7), pp.15–51.

Salingaros, N.A., 2006. *A Theory of Architecture*. Solingen: Umbau-Verlag.

Ware, C., 2013. *Information Visualization: Perception for Design*. 3rd ed. Burlington, MA: Morgan Kaufmann.

Zeki, S., 1999. *Inner Vision: An Exploration of Art and the Brain*. Oxford: Oxford University Press.

Chapter 12:

Synergy in Social Behaviour and Group Dynamics

Human beings are inherently social creatures, and much of our behaviour is shaped by the patterns we observe and mirror in the world around us. From the smallest interactions to the most complex social structures, the brain relies on its ability to recognise, mirror, and synchronise with the behaviours of others. This capacity for recursive pattern recognition not only allows individuals to function within social groups but also fosters social cohesion, cooperation, and shared identity. In this chapter, we will explore how recursive processes manifest in social behaviour, how the brain synchronises with others through shared patterns, and how these processes contribute to the formation of group dynamics and societal structures.

At the heart of human social interaction is the brain's ability to mirror the behaviour of others. This process, often referred to as social mirroring or mimicry, involves the unconscious replication of gestures, facial expressions, and vocal patterns in social situations. When we see someone smile, for example, our brain mirrors the smile by activating similar neural pathways, often leading us to smile in return. This social mirroring is not limited to facial expressions; it extends to body language, tone of voice, and even emotional states. The brain's ability to mirror these patterns allows individuals to synchronise their actions and emotions, fostering social bonds and creating a sense of shared experience.

This recursive process of mirroring and synchronisation is particularly evident in conversation. When two people engage in dialogue, their brains are constantly mirroring each other's speech patterns, body language, and emotional cues. This unconscious synchronisation ensures that the conversation flows smoothly, as each person adjusts their responses based on the patterns they observe in the other. For example, when one person raises their voice or speeds up their speech, the other person may unconsciously mirror these patterns, matching their tone and pace to maintain the rhythm of the conversation. This dynamic interplay of mirroring and adjustment allows for a seamless exchange of ideas and emotions, even when the conversation becomes complex or emotionally charged.

The brain's ability to mirror and synchronise with others also plays a central role in empathy, the capacity to understand and share the feelings of another person. Empathy relies on the brain's ability to mirror the emotional patterns of others, allowing us to feel what they are feeling. This process is supported by mirror neurons, specialised cells in the brain that activate when we observe someone else's actions or emotions, as if we were experiencing those actions or emotions ourselves. These neurons enable the brain to engage in a recursive reflection of the emotional state of others, fostering emotional resonance and understanding.

In social groups, the brain's recursive mirroring processes extend beyond individual interactions to shape group behaviour. When people come together in a group, their brains mirror not only the behaviours of those around them but also the social norms and expectations that govern the group. These shared patterns of behaviour, often referred to as social scripts, guide how individuals act in different social contexts. For example, in a formal setting such as a business meeting, individuals mirror the patterns of professionalism and decorum that are expected in that context, adjusting their body language, speech, and emotional expressions to align with the group's norms.

The recursive nature of social scripts ensures that they are flexible and adaptable. While individuals may initially mirror the group's behaviour to fit in, their unique actions and responses can introduce slight variations or dislocations in the pattern, leading to the evolution of new norms or behaviours. This process is particularly evident in cultural change, where shifts in social norms and expectations are often the result of recursive dislocations in established patterns. As individuals and groups mirror each other's behaviours and introduce new variations, these patterns evolve over time, creating dynamic social systems that are constantly adapting to new challenges and circumstances.

One of the most powerful examples of recursive patterning in social behaviour can be seen in crowd dynamics. When large groups of people gather for a common purpose, such as at a concert, protest, or sporting event, their behaviours often synchronise in ways that create a collective consciousness. This phenomenon is sometimes referred to as emergent behaviour, where the actions of individuals combine to form a larger, more complex pattern that none of the individuals could have predicted or controlled on their own. The brain's ability to mirror and synchronise with the patterns of the crowd allows individuals to feel connected to the group, leading to a sense of belonging and shared identity.

For instance, in a concert setting, the audience's rhythmic clapping, dancing, or singing along to the music creates a collective pattern that reinforces the emotional intensity of the experience. The brain mirrors the auditory and visual patterns of those around us, leading to a recursive process where the individual's behaviour synchronises with the group. This creates a feedback loop of shared emotion and action, intensifying the sense of connection to both the music and the other members of the audience.

The same recursive mirroring processes can be observed in more serious contexts, such as political protests or social movements. When individuals come together to advocate for a common cause, their behaviours, chanting, marching, holding signs, mirror the collective actions of the group, creating a unified front. This recursive synchronisation of actions and emotions reinforces the group's cohesion and solidarity, making it more effective in communicating its message and achieving its goals. The brain's ability to mirror these collective patterns allows individuals to feel part of something larger than themselves, driving social change and collective action.

The power of recursive mirroring in social behaviour is also evident in how individuals and groups organise themselves in situations where formal rules or structures are absent. As we discussed in an earlier chapter, unregulated traffic environments provide a fascinating example of how individuals instinctively synchronise their actions to avoid chaos and maintain order. In these settings, drivers rely on social cues and pattern recognition to navigate complex intersections without traffic lights or road signs. The brain mirrors the movements and behaviours of other drivers, creating a recursive loop where individuals adjust their actions in response to the collective pattern, allowing hundreds of cars to pass through a busy intersection without accidents.

This ability to spontaneously organise and synchronise behaviour is not limited to traffic; it can be observed in everyday social interactions, from navigating crowded spaces to participating in group discussions. The brain's capacity for recursive pattern recognition allows individuals to anticipate and respond to the behaviours of others, creating a dynamic and adaptive social environment where cooperation and coordination are possible without the need for explicit rules or instructions.

The concept of social synchronisation also extends to the emotional realm, where groups of people can share and amplify each other's emotions. This is particularly evident in situations of collective joy or collective grief, where the emotional patterns of individuals mirror and reinforce each other, creating a shared emotional experience. In a joyful context, such as a celebration or festival, the brain mirrors the happiness and excitement of those around us, amplifying our own emotional response. Similarly, in situations of collective grief, such as a funeral or a national tragedy, the brain mirrors the sadness and sorrow of the group, deepening our sense of empathy and emotional connection.

These shared emotional patterns are not only important for individual well-being but also play a critical role in maintaining social cohesion. By synchronising their emotional responses, individuals reinforce the bonds of trust, solidarity, and mutual understanding that hold social groups together. This recursive mirroring of emotions is particularly important in close-knit communities, where shared emotional experiences create a sense of belonging and collective identity.

Rituals and traditions are another way that recursive patterns shape social behaviour and group dynamics. Rituals, whether they are religious ceremonies, cultural festivals, or family traditions, involve the repetition of specific actions, words, and symbols that have been passed down through generations. The brain's ability to mirror these patterns ensures that they are preserved and transmitted across time, reinforcing the group's identity and values. However, just as with other social patterns, rituals are not static; they evolve through recursive dislocations, as new generations introduce subtle changes that reflect the changing social and cultural landscape.

In modern society, digital technology has created new forms of social synchronisation and group dynamics. Social media platforms, for example, allow individuals to mirror and share behaviours, emotions, and ideas on a global scale. The rapid spread of memes, hashtags, and viral content reflects the brain's capacity to recognise and replicate patterns, creating recursive loops of shared behaviour and collective action. In this digital space, the brain mirrors not only the actions of those within our immediate social circles but also those of people around the world, creating new forms of social connection and identity.

However, the recursive nature of digital synchronisation also has its challenges. The rapid spread of misinformation, groupthink, or social polarisation can occur when individuals mirror and amplify patterns of behaviour or thought without critical reflection. In these cases, the brain's tendency to synchronise with the group can lead to the reinforcement of negative patterns, highlighting the importance of cognitive dislocation, the ability to introduce new perspectives and challenge existing norms, as a tool for social adaptation and change.

In summary, the brain's recursive mirroring processes are fundamental to social behaviour and group dynamics, allowing individuals to synchronise their actions, emotions, and responses with others. Whether in conversation, crowd dynamics, or collective rituals, these shared patterns of behaviour create a sense of cohesion, trust, and belonging, reinforcing the social bonds that hold communities together. At the same time, the recursive nature of these patterns allows for flexibility and adaptation, ensuring that social systems can evolve in response to new challenges and opportunities.

In the next chapter, we will explore how these recursive processes influence moral decision-making and ethical behaviour, examining how the brain's pattern recognition systems shape our understanding of right and wrong, fairness, and justice.

References

Axelrod, R., 1984. *The Evolution of Cooperation*. New York: Basic Books.

Bandura, A., 1977. *Social Learning Theory*. Englewood Cliffs, NJ: Prentice-Hall.

Cialdini, R.B. and Goldstein, N.J., 2004. Social influence: Compliance and conformity. *Annual Review of Psychology*, 55, pp.591–621.

Dunbar, R.I.M., 1998. The social brain hypothesis. *Evolutionary Anthropology*, 6(5), pp.178–190.

Festinger, L., 1950. Informal social communication. *Psychological Review*, 57(5), pp.271–282.

Granovetter, M.S., 1973. The strength of weak ties. *American Journal of Sociology*, 78(6), pp.1360–1380.

Johnson, D.W. and Johnson, R.T., 2009. *Joining Together: Group Theory and Group Skills*. 10th ed. Boston: Pearson.

Latane, B., 1981. The psychology of social impact. *American Psychologist*, 36(4), pp.343–356.

Nowak, M.A., 2006. Five rules for the evolution of cooperation. *Science*, 314(5805), pp.1560–1563.

Sherif, M., 1966. *Group Conflict and Cooperation*. London: Routledge.

Tajfel, H. and Turner, J.C., 1979. An integrative theory of intergroup conflict. In: W.G. Austin and S. Worchel, eds. *The Social Psychology of Intergroup Relations*. Monterey, CA: Brooks/Cole, pp.33–47.

Turner, J.C., 1982. Towards a cognitive redefinition of the social group. In: H. Tajfel, ed. *Social Identity and Intergroup Relations*. Cambridge: Cambridge University Press, pp.15–40.

Chapter 13:

Recursive Patterns in Moral Decision-Making and Ethical Behaviour

The brain's ability to recognise, mirror, and refine patterns is not only central to social behaviour but also to how we understand and navigate complex questions of morality and ethics. Our sense of right and wrong, as well as our decisions about what is fair or just, emerge from the brain's recursive processes of pattern recognition, where past experiences, social norms, and emotional responses are mirrored and integrated into decision-making. In this chapter, we will explore how recursive processes shape moral judgement, how the brain applies these patterns to ethical dilemmas, and how societal norms evolve through recursive reflection on morality.

At its core, moral decision-making is a process of recognising and evaluating patterns of behaviour, both in ourselves and in others. The brain constantly mirrors the moral patterns it has learned, whether through family, culture, or personal experience, against new situations, allowing us to make decisions about what is right, wrong, fair, or unjust. These patterns are shaped by a combination of cognitive reasoning, emotional empathy, and social conditioning, all of which are processed recursively to guide moral judgement.

One of the key components of moral decision-making is the brain's ability to evaluate intentions and outcomes. When faced with a moral dilemma, the brain mirrors the patterns of the situation against stored experiences of similar dilemmas, weighing both the intentions behind an action and its potential consequences. This recursive evaluation allows the brain to determine whether an action aligns with established moral patterns or whether it represents a deviation from what is considered ethical.

For example, consider the classic moral dilemma known as the trolley problem, where an individual must decide whether to divert a runaway trolley onto a track where it will kill one person, or allow it to continue on its current path, where it will kill five people. The brain mirrors the moral patterns of utilitarian thinking, the idea that the greatest good for the greatest number should guide ethical decisions, against deontological thinking, which prioritises the sanctity of individual rights. The recursive nature of this decision-making process involves reflecting on both the intentions (saving more lives) and the outcomes (actively causing harm to one person) to arrive at a moral judgement.

This recursive evaluation is not only cognitive but also emotional. Empathy, as we discussed in previous chapters, plays a central role in moral decision-making by allowing the brain to mirror the emotional experiences of others. When faced with a moral choice, the brain reflects on how the decision will impact those involved, mirroring their potential emotional responses, such as suffering, joy, or relief, against our own emotional experiences. This empathic mirroring ensures that moral decisions are not purely rational but also guided by a sense of compassion and emotional resonance.

Moral judgement also relies on the brain's ability to recognise patterns of fairness and justice. These patterns are often learned through socialisation, where individuals are exposed to cultural norms and legal systems that define what is fair and just within a given society. The brain mirrors these societal patterns against personal experiences of fairness, creating a recursive loop where moral decisions are evaluated based on both individual and collective standards.

For example, a child who is punished unfairly for something they did not do may develop a strong emotional and cognitive understanding of fairness based on that experience. This pattern of fairness is then mirrored in future moral decisions, guiding the child's sense of justice as they grow older. Similarly, societies develop collective patterns of fairness through legal and ethical systems, which individuals mirror and internalise. These societal patterns are not static but evolve over time through recursive reflection on what constitutes justice, fairness, and human rights.

The brain's recursive evaluation of moral patterns also extends to moral growth and ethical development. As individuals encounter new experiences and challenges, their moral patterns are constantly refined and updated. This process of moral refinement often involves cognitive dislocation, where the brain mirrors a new moral situation that does not align with its existing patterns, forcing the individual to reconsider or adapt their moral framework. This is particularly evident in moral dilemmas where conflicting values or principles are at play, requiring the brain to engage in a recursive process of reflection and revision.

One powerful example of moral refinement can be seen in the evolution of attitudes toward human rights. Throughout history, societies have often mirrored patterns of exclusion, discrimination, and inequality, where certain groups were denied basic rights and freedoms. However, as new moral dislocations occurred, through social movements, intellectual debates, and empathic recognition of the suffering of others, societies began to refine their understanding of justice and equality. The brain's recursive mirroring of these evolving moral patterns led to the expansion of human rights, from the abolition of slavery to the recognition of civil rights, gender equality, and LGBTQ+ rights.

This recursive process of moral refinement is not limited to individuals or specific societies but can occur on a global scale. As we encounter new ethical challenges, such as climate change, technological advancements, or economic inequality, the brain mirrors these issues against existing moral frameworks, prompting a collective reflection on how to adapt our ethical principles to address these global concerns. The United Nations Universal Declaration of Human Rights, for example, represents a recursive synthesis of moral patterns from diverse cultures, reflecting humanity's shared commitment to dignity, freedom, and justice.

One of the most intriguing aspects of moral decision-making is the brain's ability to reconcile self-interest with altruism. While evolutionary theories often suggest that individuals are primarily driven by self-interest and survival, the brain's recursive mirroring processes enable us to transcend purely selfish motivations and act in ways that benefit others, even at personal cost. This capacity for altruism is rooted in the brain's ability to mirror the emotional and social patterns of others, creating a sense of shared humanity that compels us to help those in need, protect the vulnerable, or sacrifice for the greater good.

Reciprocal altruism, a concept in evolutionary biology, highlights how altruistic behaviour can emerge from recursive social patterns. In reciprocal altruism, individuals help others with the expectation that their kindness will be returned in the future. This creates a recursive loop where social bonds are strengthened through acts of generosity, cooperation, and trust. The brain mirrors these patterns of reciprocal altruism in both personal relationships and larger social networks, allowing for the development of cooperative societies where individuals work together for the collective good.

However, the recursive nature of moral decision-making also has its challenges. The brain's reliance on social patterns and cultural norms can sometimes lead to the reinforcement of harmful moral patterns, such as prejudice, discrimination, or intolerance. When individuals or societies mirror patterns of exclusion or violence, these behaviours can become normalised, creating recursive loops that perpetuate injustice or inequality. This is why moral dislocation, the introduction of new perspectives, experiences, or ethical frameworks, is so important for breaking negative recursive patterns and fostering moral growth.

Moral dislocations often occur when individuals are confronted with the suffering or injustice experienced by others, leading to a disruption in their existing moral framework. For example, witnessing or learning about the effects of poverty, racism, or war can introduce dislocations in the brain's moral patterns, prompting individuals to reflect on their own values and consider how they can contribute to positive social change. These moral dislocations are essential for personal and societal growth, as they challenge the status quo and encourage the refinement of ethical principles.

The recursive processes that shape moral decision-making also play a crucial role in ethical systems such as philosophy, religion, and law. These systems provide structured frameworks for evaluating moral dilemmas and guiding ethical behaviour, often through the establishment of universal principles or moral codes. However, even within these systems, recursive refinement is necessary to adapt to changing social, cultural, and technological contexts.

For instance, many religious traditions have moral codes that have evolved over time through recursive reflection on sacred texts and teachings. As societies encounter new moral challenges, such as medical ethics, environmental stewardship, or human rights, religious leaders and philosophers often engage in recursive reflection to update and reinterpret moral principles in light of contemporary issues. Similarly, legal systems are constantly evolving through recursive processes, where court rulings, legislation, and constitutional amendments reflect the ongoing refinement of justice and fairness.

In summary, the brain's recursive processes of pattern recognition, empathic mirroring, and moral dislocation are fundamental to moral decision-making and ethical behaviour. Whether we are evaluating personal dilemmas, navigating complex social issues, or refining our collective understanding of justice, these recursive processes ensure that our moral judgements are dynamic, flexible, and responsive to new challenges. As we continue to reflect on and refine our moral patterns, both as individuals and as societies, we contribute to the ongoing evolution of ethical thought and action.

In the next chapter, we will explore how these recursive processes influence creativity and problem-solving in fields such as science, technology, and innovation, examining how pattern recognition and recursive thinking drive human progress.

References

Bandura, A., 1999. Moral disengagement in the perpetration of inhumanities. *Personality and Social Psychology Review*, 3(3), pp.193–209.

Bloom, P., 2013. *Just Babies: The Origins of Good and Evil*. New York: Crown Publishers.

Damasio, A., 1994. *Descartes' Error: Emotion, Reason and the Human Brain*. London: Vintage.

Greene, J.D., 2013. *Moral Tribes: Emotion, Reason and the Gap Between Us and Them*. London: Atlantic Books.

Haidt, J., 2001. The emotional dog and its rational tail: A social intuitionist approach to moral judgment. *Psychological Review*, 108(4), pp.814–834.

Kohlberg, L., 1981. *Essays on Moral Development: The Philosophy of Moral Development*. San Francisco: Harper and Row.

Moll, J., de Oliveira-Souza, R. and Eslinger, P.J., 2003. Morals and the human brain: A working model. *NeuroReport*, 14(3), pp.299–305.

Narvaez, D. and Lapsley, D.K., 2009. Moral identity, moral functioning, and the development of moral character. *Psychology of Learning and Motivation*, 50, pp.237–274.

Rest, J.R., Narvaez, D., Bebeau, M.J. and Thoma, S.J., 1999. *Postconventional Moral Thinking*. Mahwah, NJ: Lawrence Erlbaum Associates.

Rizzolatti, G. and Sinigaglia, C., 2008. *Mirrors in the Brain: How Our Minds Share Actions and Emotions*. Oxford: Oxford University Press.

Turiel, E., 1983. *The Development of Social Knowledge: Morality and Convention*. Cambridge: Cambridge University Press.

Young, L. and Dungan, J., 2012. Where in the brain is morality? *Science*, 337(6099), pp.1179–1180.

Chapter 14:

Recursive Thinking in Creativity, Science, and Innovation

Creativity and innovation are often regarded as the engines of human progress, driving advancements in science, technology, and culture. Underpinning these breakthroughs is the brain's capacity for recursive thinking, the ability to mirror, refine, and reorganise patterns to generate novel solutions, concepts, and technologies. In this chapter, we will explore how recursive processes shape creative problem-solving, how they fuel advancements in science and innovation, and how these processes mirror the larger patterns of evolution and discovery that have propelled human development.

At its core, creative thinking involves the brain's ability to take familiar patterns and introduce dislocations or variations that lead to something new. This process, as we have seen in earlier chapters, is deeply recursive. The brain mirrors existing knowledge, reflects on it, and makes subtle changes to create novel insights or approaches. Whether in art, science, or technology, creative breakthroughs often occur when the brain dislocates familiar patterns and reassembles them in unexpected ways.

One of the most famous examples of recursive creativity in science is Albert Einstein's development of the theory of relativity. Einstein's process of discovery was not purely mathematical; it involved thought experiments where he imagined scenarios, such as riding alongside a beam of light, that allowed him to mirror and refine the known patterns of physics. By recursively reflecting on these imagined scenarios and introducing dislocations to Newtonian concepts of time and space, Einstein was able to formulate a radically new understanding of the universe. His ability to mirror both abstract concepts and physical realities through recursive thinking highlights the brain's capacity to generate innovative ideas by breaking and reconfiguring familiar patterns.

Similarly, Isaac Newton's discovery of calculus emerged from recursive thinking, as he mirrored and reflected on patterns of motion and change that had been observed but not mathematically formalised. Newton's ability to mirror these natural patterns and introduce a new mathematical framework for describing them illustrates the recursive nature of scientific discovery. Each time a new layer of understanding was added, Newton recursively refined his models, leading to a breakthrough that would fundamentally alter the course of science.

In the realm of technology, recursive thinking plays a critical role in engineering and design. Innovators often build on existing technologies, mirroring the patterns of past designs while introducing refinements that address current challenges. This recursive process of iteration, where a prototype is tested, refined, and improved, mirrors the brain's ability to recognise and adjust patterns based on feedback. The recursive loop between design, testing, and refinement is what drives technological advancements, ensuring that each new generation of technology is an improvement on the last.

Steve Jobs and Apple's innovations in personal computing and mobile technology are prime examples of recursive thinking in design. Jobs' vision for devices like the iPhone was rooted in mirroring the patterns of existing technologies, such as computers, phones, and media players, but introducing dislocations that led to a revolutionary new product. By recursively refining the user interface, aesthetics, and functionality, Jobs and his team were able to create a device that not only met but exceeded users' expectations. This recursive process of refinement is what allowed Apple to continue innovating, pushing the boundaries of what technology could achieve.

Problem-solving, whether in science, technology, or everyday life, relies on recursive thinking. When faced with a complex problem, the brain mirrors the problem's structure against patterns it has encountered before. It then recursively refines these patterns, introducing variations or testing new approaches until a solution emerges. This process is not linear but involves constant feedback loops, where each new attempt introduces slight dislocations that gradually refine the solution.

In mathematics and engineering, problem-solving often follows this recursive pattern of testing and refining hypotheses or models. For example, Leonardo da Vinci's notebooks are filled with sketches of inventions, such as flying machines and hydraulic systems, where he mirrored patterns observed in nature and refined them into technological designs. Each sketch represented a recursive iteration of his thinking, where he introduced slight changes to the design based on his observations of birds in flight or the flow of water. Although many of his inventions were never built during his lifetime, da Vinci's recursive thinking laid the groundwork for future innovations.

Evolutionary biology provides another powerful analogy for recursive thinking in creativity and innovation. Just as biological evolution relies on recursive processes, where genetic patterns are replicated, dislocated through mutation, and refined through natural selection, human creativity involves the recursive refinement of ideas, where dislocations lead to novel adaptations. In both evolution and innovation, recursive feedback loops ensure that patterns are continuously refined and adapted to changing environments.

The recursive processes that drive creativity and problem-solving are not limited to individuals; they are also evident in collective innovation and scientific collaboration. In fields such as medicine, physics, and space exploration, breakthroughs often emerge from the recursive mirroring and refinement of ideas across teams, institutions, and generations. For example, the development of the Human Genome Project, which mapped the entire sequence of human DNA, involved recursive collaboration between scientists from around the world. Each stage of the project mirrored previous discoveries in genetics, while introducing new technologies and techniques that refined the understanding of the human genome.

Open-source software development is another example of recursive collaboration, where programmers around the world mirror and refine each other's code, introducing new features, fixing bugs, and adapting the software to meet evolving needs. This recursive loop of coding, testing, and refinement ensures that open-source projects are dynamic and continuously improving, much like the brain's recursive processes of pattern recognition and refinement.

The recursive nature of scientific and technological innovation is also reflected in the scientific method itself, which relies on recursive loops of hypothesis, experiment, and revision. When scientists develop a hypothesis, they mirror existing knowledge and refine it into a testable prediction. The results of the experiment are then mirrored against the hypothesis, leading to recursive adjustments or refinements. This recursive feedback loop between theory and experimentation ensures that scientific knowledge is constantly evolving, adapting to new evidence and insights.

In creative industries such as film, music, and literature, recursive thinking plays a central role in artistic innovation. Just as in scientific fields, artists and creators mirror existing patterns, whether they be narrative structures, musical forms, or visual styles, and introduce dislocations to create something new. In cinema, directors like Stanley Kubrick and Christopher Nolan are known for their recursive storytelling techniques, where narrative patterns are mirrored, layered, and reflected back upon themselves, creating complex, multi-dimensional films that challenge viewers to engage in recursive thinking.

Music, as we explored in a previous chapter, also thrives on recursive patterns, where familiar melodies, rhythms, and harmonies are mirrored and then dislocated through variations to create emotional resonance and novelty. Jazz improvisation, for instance, involves musicians recursively mirroring and dislocating musical patterns in real time, creating spontaneous and innovative performances that push the boundaries of musical expression.

In literature, authors like James Joyce, Virginia Woolf, and Italo Calvino use recursive narrative structures to create works that mirror the complexity of human thought and experience. Joyce's *Ulysses*, for example, mirrors the structure of Homer's Odyssey, while dislocating and reframing the epic narrative within the context of modern life. This recursive layering of classical and contemporary patterns allows Joyce to explore themes of identity, memory, and consciousness in new and profound ways.

One of the most exciting frontiers of recursive thinking in science and innovation is the field of artificial intelligence (AI) and machine learning. Machine learning algorithms are, in essence, recursive systems that mirror patterns in data, refine those patterns through feedback, and generate increasingly accurate predictions or solutions. These algorithms rely on recursive loops of training and testing, where each iteration introduces refinements based on the data processed. The recursive nature of machine learning has already led to breakthroughs in fields ranging from healthcare to autonomous vehicles, and it holds the potential to revolutionise how we solve problems and create new technologies.

However, as with all recursive systems, the dislocations introduced by AI and machine learning must be carefully managed to ensure that the patterns being mirrored and refined lead to positive outcomes. The ethical implications of recursive AI systems, such as bias in data sets or the unintended consequences of autonomous decision-making, require recursive reflection and refinement of both the technology and the ethical frameworks that govern its use.

In summary, recursive thinking is the driving force behind creativity, scientific discovery, and technological innovation. Whether in the mind of an individual artist or scientist, or in the collective efforts of teams and societies, the recursive mirroring, refining, and dislocating of patterns is what allows humans to push the boundaries of knowledge and create new possibilities. As we continue to refine our understanding of recursive thinking, we will unlock new ways to harness this powerful process for solving the complex challenges of the future.

In the next chapter, we will explore how recursive thinking shapes our understanding of time, memory, and the self, examining how the brain's recursive processes allow us to construct a continuous sense of identity and navigate the temporal world.

References

Boyd, R. and Richerson, P.J., 1985. *Culture and the Evolutionary Process*. Chicago: University of Chicago Press.

Cavalli-Sforza, L.L. and Feldman, M.W., 1981. *Cultural Transmission and Evolution: A Quantitative Approach*. Princeton: Princeton University Press.

Dawkins, R., 1976. *The Selfish Gene*. Oxford: Oxford University Press.

Henrich, J., 2016. *The Secret of Our Success: How Culture Is Driving Human Evolution*. Princeton: Princeton University Press.

Mesoudi, A., 2011. *Cultural Evolution: How Darwinian Theory Can Explain Human Culture and Synthesize the Social Sciences*. Chicago: University of Chicago Press.

Rogers, E.M., 2003. *Diffusion of Innovations*. 5th ed. New York: Free Press.

Shennan, S., 2002. *Genes, Memes and Human History: Darwinian Archaeology and Cultural Evolution*. London: Thames and Hudson.

Sperber, D., 1996. *Explaining Culture: A Naturalistic Approach*. Oxford: Blackwell.

Tomasello, M., 1999. *The Cultural Origins of Human Cognition*. Cambridge, MA: Harvard University Press.

Whiten, A., Horner, V. and de Waal, F.B.M., 2005. Conformity to cultural norms of tool use in chimpanzees. *Nature*, 437(7059), pp.737–740.

Chapter 15:

Recursive Patterns in Time, Memory, and the Self

Our understanding of time, memory, and the self is deeply rooted in the brain's capacity for recursive pattern recognition and reflection. The brain constantly mirrors past experiences against the present and anticipates future possibilities, creating a seamless flow that allows us to construct a continuous sense of identity and navigate the temporal world. In this chapter, we will explore how recursive processes shape our perception of time, how they influence the formation and retrieval of memory, and how they contribute to our evolving sense of self.

The brain's ability to recognise and reflect on temporal patterns is essential for understanding the flow of time. Time, as we perceive it, is not a series of disconnected moments but a continuous stream that the brain organises into patterns of past, present, and future. This organisation is made possible by the brain's recursive ability to mirror past experiences, compare them to present conditions, and project future outcomes.

One of the most fundamental ways the brain interacts with time is through temporal memory, where past experiences are stored, mirrored, and reflected upon in the present. Memory is inherently recursive: each time we recall a memory, the brain mirrors that pattern of the past and overlays it with the present context, introducing subtle dislocations or updates. This recursive reflection ensures that memories are not static but are continually refined, evolving as we integrate new experiences and insights.

For example, a person reflecting on a childhood event will recall the sensory details, emotions, and thoughts associated with that moment. However, as the brain mirrors that memory in the present, it may introduce new layers of meaning based on the individual's current understanding of life. This recursive process explains why our memories can change over time and why certain events take on new significance as we gain new perspectives. The brain's ability to reflect on past experiences and adjust the emotional or cognitive meaning of those memories is crucial for personal growth and adaptation.

The recursive nature of memory is also evident in how the brain consolidates information during sleep. During sleep, the brain mirrors and reorganises patterns of experiences from the day, refining them into long-term memory. This process of memory consolidation is a form of recursive mirroring, where the brain revisits and strengthens neural connections associated with important experiences while pruning those that are less relevant. In this way, the brain's recursive processes ensure that memories are not only preserved but also refined and adapted to support learning and future behaviour.

Time perception itself is a deeply recursive process. The brain does not passively experience time as a linear sequence; instead, it actively constructs a sense of temporal continuity by mirroring and layering patterns of events. This recursive construction allows us to perceive time in a flexible and adaptive way, where we can compress or expand time based on context. For example, when we are deeply engaged in an activity, time may seem to fly by, as the brain mirrors the repetitive patterns of the task in a way that reduces our perception of the passage of time. Conversely, during moments of boredom or discomfort, time may seem to drag, as the brain mirrors the lack of stimulating patterns, creating a sense of temporal elongation.

This ability to stretch and compress time perception is essential for navigating the world efficiently. By mirroring and adjusting temporal patterns, the brain allows us to focus on what is important in the present while minimising the distractions of time's passage. In high-pressure situations, such as during a competitive sporting event or a life-threatening situation, the brain's recursive processes enable us to slow down our perception of time, giving us the cognitive space to make quick decisions and react appropriately. This temporal flexibility is a key aspect of how the brain optimises our interaction with the environment.

The brain's recursive processes also play a central role in how we construct our sense of self. Our identity is not a fixed entity but a dynamic, evolving narrative that the brain constructs through the recursive reflection of past experiences, current thoughts, and future aspirations. Each time we reflect on who we are, the brain mirrors patterns of self-concept, comparing them against both past identities and future goals. This recursive process allows us to maintain a coherent sense of self over time, even as we undergo significant changes in our beliefs, behaviours, and life circumstances.

For example, an individual may think of themselves as a compassionate person, a pattern of self-concept that has been reinforced by past actions and experiences. However, when faced with a situation that challenges this self-concept, such as a moral dilemma or personal failure, the brain mirrors this new experience against the established pattern of identity. The resulting dislocation may prompt the individual to refine or adjust their self-concept, leading to personal growth or a reevaluation of their values. This recursive refinement of self-concept is what allows individuals to adapt to new experiences while maintaining a sense of continuity in their identity.

The brain's recursive construction of the self is also influenced by social interactions and cultural patterns. As social beings, we constantly mirror the patterns of behaviour, values, and expectations of those around us, integrating these external patterns into our sense of self. This process of social mirroring is recursive, as we reflect on how others perceive us and adjust our behaviour or identity accordingly. The brain's ability to recursively mirror and integrate social feedback allows us to navigate complex social environments, where our identity is shaped by both internal reflections and external influences.

In literature and philosophy, the concept of the self as a recursive construction has been explored through narratives that reflect the fluid and evolving nature of identity. Marcel Proust's *In Search of Lost Time*, for example, is a literary exploration of how memory and time shape the self. The protagonist's reflections on his past, mirrored recursively throughout the novel, highlight how the self is constantly being reconstructed through the interplay of memory, emotion, and time. Proust's recursive narrative structure mirrors the brain's own processes of self-reflection, where the past is continually revisited and reinterpreted in light of present experiences.

In psychology, the concept of the narrative self has gained prominence, suggesting that individuals construct their identity through the stories they tell about themselves. These personal narratives are recursive, as they involve reflecting on past events, integrating them into a coherent story, and adjusting that story as new experiences emerge. The brain's recursive ability to mirror and refine these self-narratives allows us to make sense of our lives, giving meaning to our experiences and creating a sense of purpose and direction.

The recursive construction of the self is also influenced by future-oriented thinking. Just as the brain mirrors past experiences in memory, it also anticipates future possibilities by projecting patterns of behaviour and outcomes into the future. This ability to simulate the future is a key aspect of planning, goal-setting, and personal growth. When we imagine future scenarios, whether positive or negative, the brain mirrors these projections against our current self-concept, allowing us to adjust our actions and decisions to align with our future goals.

For instance, a person who envisions themselves achieving a major career milestone may mirror that future self against their present actions, motivating them to take steps toward their goal. Conversely, the brain may simulate negative outcomes, such as failure or regret, prompting the individual to avoid certain behaviours or reconsider their path. This recursive projection of the self into the future ensures that our actions are not only guided by past experiences but also by future aspirations, creating a dynamic feedback loop that shapes our identity over time.

Trauma and recovery provide a powerful example of how the brain's recursive processes can both disrupt and restore the self. In the case of trauma, the brain mirrors the traumatic event repeatedly, often leading to intrusive thoughts or flashbacks that can distort the individual's sense of time and identity. This recursive mirroring can make it difficult for the brain to move beyond the trauma, as the patterns of fear, pain, and loss are continuously mirrored against the present. However, through processes like therapy or personal reflection, individuals can gradually introduce new dislocations to these traumatic patterns, allowing for the integration of the trauma into a larger, more coherent self-narrative. This recursive process of healing enables individuals to reclaim their sense of self and reestablish a sense of temporal continuity.

In summary, the brain's recursive processes are essential for our understanding of time, memory, and the self. Through recursive reflection and refinement, the brain constructs a continuous narrative of identity that integrates past experiences, present realities, and future aspirations. These recursive loops ensure that we are able to navigate the temporal world with a sense of coherence, adaptability, and growth. Whether through the mirroring of memories, the anticipation of future goals, or the recursive refinement of self-concept, the brain's ability to recognise and refine patterns is what allows us to maintain a stable yet evolving sense of who we are.

In the next chapter, we will explore how these recursive processes shape our understanding of consciousness, examining the relationship between recursive thinking and our awareness of the world, others, and ourselves.

References

Argyris, C. and Schön, D.A., 1978. *Organizational Learning: A Theory of Action Perspective*. Reading, MA: Addison-Wesley.

Campbell, D.T., 1960. Blind variation and selective retention in creative thought as in other knowledge processes. *Psychological Review*, 67(6), pp.380–400.

Donald, M., 1991. *Origins of the Modern Mind*. Cambridge, MA: Harvard University Press.

Kuhn, T.S., 1962. *The Structure of Scientific Revolutions*. Chicago: University of Chicago Press.

Nonaka, I. and Takeuchi, H., 1995. *The Knowledge-Creating Company*. Oxford: Oxford University Press.

Piaget, J., 1970. *Genetic Epistemology*. New York: Columbia University Press.

Popper, K., 1972. *Objective Knowledge: An Evolutionary Approach*. Oxford: Oxford University Press.

Schön, D.A., 1983. *The Reflective Practitioner*. New York: Basic Books.

Sterelny, K., 2012. *The Evolved Apprentice*. Cambridge, MA: MIT Press.

Toulmin, S., 1972. *Human Understanding: The Collective Use and Evolution of Concepts*. Oxford: Oxford University Press.

Vygotsky, L.S., 1978. *Mind in Society*. Cambridge, MA: Harvard University Press.

Chapter 16:

Recursive Patterns in Consciousness and Self-Awareness

Consciousness, the state of being aware of oneself and the world, remains one of the most complex and debated topics in neuroscience and philosophy. At its core, consciousness involves the brain's recursive ability to mirror, process, and reflect on both external stimuli and internal states. Through recursive patterns of self-awareness and cognitive reflection, the brain creates a continuous sense of subjective experience. In this chapter, we will explore how recursive processes shape consciousness, how they contribute to self-awareness, and how these processes allow us to reflect on our thoughts, emotions, and existence.

Consciousness can be understood as a multi-layered recursive system where the brain mirrors and integrates sensory inputs, cognitive processes, and emotional states to produce a unified experience. This recursive layering allows us to be aware of our surroundings, process incoming information, and reflect on our own thoughts and feelings. While basic consciousness involves sensory awareness, higher-order consciousness, the ability to reflect on one's own thoughts and experiences, requires more complex recursive processing, where the brain mirrors not only external stimuli but also its own cognitive operations.

One of the key components of consciousness is the brain's ability to generate self-awareness. Self-awareness involves the recursive mirroring of the self as an object of reflection, where the brain reflects on its own thoughts, emotions, and actions. This recursive self-reflection is what allows us to experience introspection, the ability to look inward and evaluate our own mental states. For example, when we feel anxious, we can reflect on that anxiety, analyse its causes, and attempt to regulate it. This recursive loop of self-monitoring and self-regulation is a central feature of conscious experience.

Neuroscientific research suggests that self-awareness is linked to specific areas of the brain, particularly the prefrontal cortex, which is involved in decision-making, planning, and self-regulation. The prefrontal cortex plays a critical role in executive function, enabling the brain to recursively reflect on goals, evaluate progress, and adjust behaviour accordingly. This recursive monitoring of actions and thoughts is essential for maintaining a coherent sense of self across time.

The brain's recursive ability to reflect on its own operations is also linked to the concept of metacognition, thinking about thinking. Metacognition involves the brain's capacity to monitor its cognitive processes, assess the effectiveness of those processes, and make adjustments when necessary. For example, when solving a problem, we may reflect on whether our current approach is working or whether we need to change strategies. This recursive monitoring of cognitive performance allows us to adapt and optimise our thinking in real time, enhancing both learning and decision-making.

In philosophy of mind, the recursive nature of consciousness has been explored in the context of self-referential awareness, where the mind is seen as recursively reflecting on its own states and operations. This self-referential loop is what gives rise to subjective experience or qualia, the felt quality of experience. The fact that we can experience not only sensations but also the awareness of those sensations, such as the warmth of the sun or the sweetness of chocolate, suggests that consciousness involves multiple layers of recursive reflection. Each layer of awareness builds on the previous one, creating a richer, more nuanced experience of the world.

David Hume, the 18th-century philosopher, famously questioned whether the self exists as a stable entity or is merely a collection of fleeting perceptions. Hume's scepticism about the self-highlights the recursive nature of consciousness: the self is not a fixed object but an ongoing process of recursive reflection, where the brain mirrors its own experiences and constructs a sense of continuity. This continuous mirroring creates the illusion of a stable self, even though our thoughts, emotions, and perceptions are constantly changing.

The recursive nature of consciousness is also evident in how we experience time and continuity. As discussed in the previous chapter, the brain mirrors past experiences, compares them to the present, and anticipates the future. This recursive process allows us to maintain a continuous sense of self across time, even though our thoughts and experiences may vary from moment to moment. For example, when we wake up in the morning, we immediately re-establish a sense of who we are based on recursive reflections of past experiences and future expectations. This ability to maintain a continuous sense of self despite the passage of time is a key feature of conscious awareness.

The brain's recursive processes also contribute to our experience of agency, the sense that we are in control of our actions and decisions. When we decide to move our hand, for example, the brain mirrors the intention to move against the actual movement, creating a recursive loop of action and feedback. This recursive mirroring allows us to experience ourselves as agents who can initiate and control actions in the world. The sense of agency is closely tied to our experience of free will, the belief that we can make choices and direct our own behaviour. Whether free will is an illusion or a genuine aspect of human experience remains a subject of philosophical debate, but it is clear that the brain's recursive processes are essential for generating the experience of agency.

Lucid dreaming provides a fascinating example of recursive consciousness in action. During lucid dreaming, individuals become aware that they are dreaming and can reflect on the content of the dream while it is happening. This recursive awareness of being in a dream allows for greater control over the dream narrative, as the brain mirrors the dream state and introduces conscious adjustments. Lucid dreaming demonstrates the brain's capacity to create recursive loops of awareness even within altered states of consciousness, where normal distinctions between reality and imagination become blurred.

In meditation and mindfulness practices, recursive reflection plays a central role in enhancing conscious awareness. Through meditation, individuals are trained to become more aware of their thoughts, emotions, and sensations in the present moment, without judgement. This heightened state of self-awareness involves recursive monitoring, where the brain reflects on its own mental states and patterns of attention. By engaging in this recursive practice, meditators can develop greater control over their mental processes, reduce stress, and cultivate a deeper sense of inner peace.

The recursive nature of consciousness is also reflected in creative thought, where the brain mirrors ideas, refines them, and introduces novel connections. For example, when an artist is engaged in the creative process, they may reflect on their initial concept, make adjustments, and refine their vision recursively until the artwork reaches its final form. This recursive loop of creation and reflection is what allows artists, writers, and thinkers to generate original ideas and push the boundaries of human expression.

Social consciousness, the awareness of others and our place within a social context, also relies on recursive thinking. The brain mirrors the behaviour and emotions of others through empathy, allowing us to understand their perspectives and anticipate their reactions. This recursive reflection on social dynamics enables us to navigate complex interpersonal relationships and contribute to the social fabric. The brain's recursive processes allow for the development of theory of mind, the ability to understand that others have thoughts, desires, and beliefs different from our own. This recursive awareness is critical for social interactions, cooperation, and moral decision-making.

However, the recursive nature of consciousness also introduces certain challenges, particularly when it comes to self-reflection. While the ability to reflect on our own thoughts and emotions is essential for self-awareness, it can also lead to rumination or overthinking. In some cases, individuals may become trapped in recursive loops of negative thought, where they continually reflect on past mistakes, failures, or anxieties without finding resolution. This recursive mirroring of negative patterns can contribute to conditions such as depression or anxiety, highlighting the importance of balance in recursive reflection.

The recursive processes that underlie consciousness also raise important questions about the nature of artificial intelligence and machine consciousness. If consciousness arises from recursive reflection on sensory inputs and internal states, could an AI system that mirrors its own operations achieve a form of awareness? While current AI systems are capable of recursive processing, such as in machine learning algorithms, they do not possess the self-referential awareness that characterises human consciousness. However, the exploration of recursive processes in AI offers a fascinating avenue for understanding the mechanisms that give rise to subjective experience and self-awareness.

In summary, the brain's recursive processes are fundamental to the experience of consciousness and self-awareness. Through recursive mirroring and reflection, the brain creates a continuous sense of subjective experience, allowing us to reflect on our thoughts, emotions, and actions. This recursive self-reflection is what gives rise to introspection, agency, and the ability to maintain a coherent sense of self across time. Whether in lucid dreaming, meditation, or creative thought, the recursive nature of consciousness allows us to explore the depths of our inner world and navigate the complexities of our outer world.

In the next chapter, we will examine how recursive processes shape learning and education, exploring how the brain's ability to mirror and refine patterns contributes to cognitive development, skill acquisition, and the transmission of knowledge across generations.

References

Clark, A., 2013. Whatever next? Predictive brains, situated agents, and the future of cognitive science. *Behavioral and Brain Sciences*, 36(3), pp.181–204.

Gigerenzer, G., 2007. *Gut Feelings: The Intelligence of the Unconscious*. London: Penguin Books.

Holland, J.H., 1992. *Adaptation in Natural and Artificial Systems*. Cambridge, MA: MIT Press.

Kahneman, D. and Tversky, A., 1979. Prospect theory: An analysis of decision under risk. *Econometrica*, 47(2), pp.263–292.

Kirsh, D., 1991. Today the earwig, tomorrow man? *Artificial Intelligence*, 47(1–3), pp.161–184.

Miller, G.A., Galanter, E. and Pribram, K.H., 1960. *Plans and the Structure of Behavior*. New York: Holt, Rinehart and Winston.

Rosen, R., 1985. *Anticipatory Systems: Philosophical, Mathematical and Methodological Foundations*. Oxford: Pergamon Press.

Simon, H.A., 1996. *The Sciences of the Artificial*. 3rd ed. Cambridge, MA: MIT Press.

Sterman, J.D., 2000. *Business Dynamics: Systems Thinking and Modeling for a Complex World*. Boston: McGraw-Hill.

Taleb, N.N., 2007. *The Black Swan*. London: Penguin Books.

Chapter 17:

Recursive Processes in Learning and Education

Learning is one of the most fundamental aspects of human development, and it is deeply shaped by the brain's ability to recognise, mirror, and refine patterns through recursive processes. Whether we are acquiring new skills, absorbing knowledge, or solving complex problems, the brain's recursive capacity for reflection and refinement allows us to continuously build on past experiences and adapt to new challenges. In this chapter, we will explore how recursive processes underlie learning, how they shape the acquisition of cognitive skills, and how they influence the way knowledge is transmitted in educational systems.

At its core, learning involves the brain's ability to identify patterns in the world and mirror them against previous experiences. This process is recursive because each time the brain encounters new information, it reflects on what it already knows and adjusts its understanding accordingly. This dynamic interplay between past knowledge and new input allows for cognitive growth and the formation of new neural connections.

One of the foundational mechanisms for learning is neuroplasticity, the brain's ability to reorganise itself by forming new synaptic connections. Neuroplasticity is inherently recursive because it involves the brain mirroring patterns of activity in response to repeated experiences, refining those patterns to strengthen certain connections while pruning others. When we learn a new skill, such as playing a musical instrument or speaking a new language, our brain creates neural circuits that mirror the patterns of that skill. Over time, through practice and repetition, the brain recursively refines those circuits, making the skill more automatic and efficient.

In early childhood development, recursive processes are particularly evident as children rapidly acquire language, motor skills, and social knowledge. For instance, language learning involves the recursive mirroring of phonetic patterns, syntactic rules, and vocabulary. Children listen to speech, mirror the sounds they hear, and then refine their ability to produce those sounds through recursive feedback. This process continues as they learn more complex linguistic structures, such as grammar and syntax, recursively building their ability to communicate effectively.

The brain's recursive ability to mirror and refine patterns also plays a key role in mathematical learning. When students first encounter basic mathematical concepts, such as addition and subtraction, the brain mirrors these new patterns against existing knowledge, refining its understanding through repetition and practice. As students progress to more complex concepts, such as algebra or calculus, the recursive process of reflection and refinement allows them to build on foundational knowledge and apply it to more abstract problems. This ability to recursively layer new information onto existing patterns is essential for the development of mathematical reasoning.

In problem-solving, recursive processes enable the brain to evaluate multiple possible solutions by mirroring different approaches and refining them based on feedback. When faced with a complex problem, the brain may initially mirror past solutions that have worked in similar situations, but as it tests these solutions and gathers new information, it recursively refines its approach. This trial-and-error process is recursive because each iteration introduces new dislocations or variations, leading to a refined understanding of the problem and potential solutions.

This recursive approach to problem-solving is particularly evident in scientific inquiry, where hypotheses are tested, refined, and tested again. The scientific method itself is inherently recursive, involving a continuous loop of observation, hypothesis formation, experimentation, and revision. Each time a hypothesis is tested, the results are mirrored against the initial predictions, and the hypothesis is refined based on the feedback. This recursive loop ensures that scientific knowledge evolves over time, with each iteration bringing us closer to a more accurate understanding of the natural world.

In education, recursive processes are integral to effective teaching and learning. Constructivist approaches to education, for instance, emphasise the importance of building new knowledge on top of existing understanding, mirroring the brain's natural recursive learning processes. Students are encouraged to reflect on what they already know, integrate new information, and refine their understanding through active learning and problem-solving. This recursive approach to education helps students develop deeper and more meaningful connections with the material, rather than simply memorising facts.

Scaffolding is a key teaching strategy that mirrors the brain's recursive learning processes. In scaffolding, teachers provide support that helps students navigate new concepts, gradually removing that support as students gain confidence and mastery. This process allows students to recursively refine their understanding, building on their knowledge step by step. For example, in learning to write essays, students may first receive support in structuring their arguments, but as they gain more practice, they are able to write more independently, recursively refining their writing skills over time.

Feedback loops are another important aspect of recursive learning in education. When students receive feedback on their performance, whether through grades, comments, or peer review, the brain mirrors this feedback against its own expectations, leading to recursive adjustments in behaviour or understanding. Feedback allows students to identify areas of improvement and refine their approach, creating a continuous cycle of learning and growth. This recursive loop of performance, feedback, and adjustment is essential for mastery of any skill or subject.

In collaborative learning environments, recursive processes are amplified as students mirror and refine each other's ideas through discussion, debate, and group work. When students engage in collaborative activities, they are exposed to multiple perspectives and approaches, allowing them to recursively reflect on their own understanding and adjust their thinking. This collective recursive process not only enhances individual learning but also fosters a deeper sense of community and shared knowledge.

Project-based learning is a teaching approach that leverages recursive processes to engage students in deep, meaningful learning experiences. In project-based learning, students work on complex, real-world problems that require them to recursively gather information, test ideas, and refine their solutions over time. This approach mirrors the recursive nature of problem-solving in the real world, where solutions are not reached in a single step but are the result of continuous reflection and adjustment. Project-based learning encourages students to think critically, collaborate with others, and develop the persistence needed to navigate challenges.

In lifelong learning, the brain's recursive processes ensure that individuals continue to build on their knowledge and experiences throughout their lives. As we encounter new situations and challenges, the brain mirrors these experiences against past knowledge, allowing us to adapt and grow. This recursive ability to reflect, refine, and integrate new information is what enables us to remain flexible and open to learning, even as we age.

The role of recursive processes in cultural transmission is also significant. Cultural knowledge, whether it be in the form of traditions, language, or social norms, is passed down from one generation to the next through recursive mirroring and refinement. Children learn cultural practices by observing and mirroring the behaviours of their parents and community members, recursively refining their understanding as they grow. Over time, cultural patterns evolve through recursive dislocations, as each generation introduces new variations or adaptations in response to changing circumstances. This recursive process ensures that cultural knowledge is preserved while remaining flexible and open to innovation.

In modern education, digital technologies and online learning platforms have introduced new forms of recursive learning. Adaptive learning technologies, for example, use algorithms that mirror students' performance and adjust the content accordingly. These platforms provide tailored feedback and adjust the difficulty of tasks based on the learner's progress, creating a recursive learning loop that adapts to the individual's needs. By mirroring the brain's natural recursive learning processes, these technologies have the potential to personalise education and enhance learning outcomes.

However, the recursive nature of learning also presents challenges, particularly in the context of standardised testing and rote memorisation. When education systems focus too heavily on the memorisation of facts without encouraging recursive reflection and problem-solving, students may struggle to develop the deep, flexible understanding needed for lifelong learning. In contrast, educational approaches that emphasise critical thinking, creativity, and reflection are more aligned with the brain's recursive processes, promoting deeper engagement and mastery of the material.

In summary, learning is a deeply recursive process, where the brain continuously mirrors new information against existing knowledge, refines its understanding, and adapts to new challenges. This recursive ability underpins not only individual cognitive development but also the transmission of knowledge across generations and cultures. By recognising the role of recursive processes in learning, educators can design more effective teaching strategies that align with the brain's natural capacity for reflection, refinement, and growth.

In the next chapter, we will explore how recursive processes shape human relationships and social dynamics, focusing on how the brain mirrors and refines patterns of interaction to create deep emotional connections and navigate complex social environments.

References

Bishop, C.M., 2006. *Pattern Recognition and Machine Learning.* New York: Springer.

Clark, A. and Chalmers, D., 1998. The extended mind. *Analysis*, 58(1), pp.7–19.

Hassabis, D., Kumaran, D., Summerfield, C. and Botvinick, M., 2017. Neuroscience-inspired artificial intelligence. *Neuron*, 95(2), pp.245–258.

Holland, J.H., 1995. *Hidden Order: How Adaptation Builds Complexity.* Reading, MA: Addison-Wesley.

Lake, B.M., Ullman, T.D., Tenenbaum, J.B. and Gershman, S.J., 2017. Building machines that learn and think like people. *Behavioral and Brain Sciences*, 40, pp.1–58.

LeCun, Y., Bengio, Y. and Hinton, G., 2015. Deep learning. *Nature*, 521(7553), pp.436–444.

Newell, A. and Simon, H.A., 1976. Computer science as empirical inquiry. *Communications of the ACM*, 19(3), pp.113–126.

Rumelhart, D.E., Hinton, G.E. and Williams, R.J., 1986. Learning representations by back-propagating errors. *Nature*, 323(6088), pp.533–536.

Russell, S. and Norvig, P., 2016. *Artificial Intelligence: A Modern Approach*. 3rd ed. Harlow: Pearson.

Sutton, R.S. and Barto, A.G., 2018. *Reinforcement Learning: An Introduction*. 2nd ed. Cambridge, MA: MIT Press.

Wolpert, D.M. and Flanagan, J.R., 2016. Computations underlying sensorimotor learning. *Current Opinion in Neurobiology*, 37, pp.7–11.

Chapter 18:

Recursive Patterns in Human Relationships and Social Dynamics

Human relationships are built on the brain's ability to recognise, mirror, and refine social patterns, allowing us to form emotional connections, navigate complex interactions, and build social structures. The recursive processes that shape our understanding of ourselves also play a central role in how we engage with others, helping us to predict behaviour, communicate effectively, and foster empathy. In this chapter, we will explore how recursive processes influence interpersonal relationships, how they shape group dynamics, and how they contribute to the formation of social bonds and cultural identities.

At the foundation of human relationships is the brain's ability to engage in social mirroring, where we unconsciously mirror the gestures, facial expressions, and emotions of those around us. This process of empathic mirroring allows us to synchronise our emotional and behavioural states with others, creating a sense of shared experience. For example, when we see someone smile, our brain mirrors their expression, often prompting us to smile in return. This mirroring is recursive, as the brain reflects and adjusts its own emotional state in response to the feedback it receives from others, creating a loop of emotional resonance.

This recursive emotional mirroring is central to the development of empathy, the ability to understand and share the feelings of others. When we see someone in distress, our brain mirrors their emotional state, allowing us to feel what they are feeling. This recursive process enables us to respond compassionately, fostering emotional connections and deepening social bonds. The ability to recursively mirror the emotional states of others is a fundamental aspect of human sociality, as it allows us to build relationships based on mutual understanding and care.

Nonverbal communication, such as body language and facial expressions, is another area where recursive processes shape human relationships. When we interact with others, the brain continuously mirrors and adjusts to their nonverbal cues, refining its understanding of their emotional state and intentions. This recursive loop of nonverbal communication allows us to engage in subtle social negotiations, where we adjust our own behaviour in response to the feedback we receive from others. For example, in a conversation, we may lean forward or make eye contact to signal interest, and as the other person mirrors these cues, the conversation becomes more engaging and dynamic.

In romantic relationships, recursive processes play a significant role in the development of intimacy and attachment. As partners engage in emotional and physical exchanges, their brains mirror each other's behaviours and emotions, creating a recursive loop of mutual reinforcement. Over time, this recursive mirroring deepens the emotional bond between partners, leading to a sense of shared identity and connection. The brain's ability to recursively reflect on past interactions also allows couples to navigate challenges and resolve conflicts, as they refine their understanding of each other's needs and emotional responses.

The recursive nature of relationships is not limited to individual interactions but extends to group dynamics. In social groups, individuals mirror the behaviours, attitudes, and values of the group, creating a recursive loop of social reinforcement. This process of social synchronisation ensures that group members align their behaviour with the group's norms and expectations, promoting cohesion and cooperation. For example, in a work setting, employees may mirror the leadership style and work habits of their colleagues, recursively adjusting their own behaviour to fit the group dynamic.

Social identity is also shaped by recursive processes, as individuals mirror the patterns of their social and cultural groups, integrating these patterns into their sense of self. For instance, a person's identity as a member of a particular ethnic group, religious community, or political movement is reinforced through the recursive reflection of shared values, beliefs, and traditions. This recursive mirroring of social and cultural patterns creates a sense of belonging, as individuals align themselves with the larger group and contribute to the collective identity.

One of the most powerful examples of recursive social dynamics can be seen in the phenomenon of groupthink, where the recursive reinforcement of group norms and ideas can lead to a narrowing of perspectives. In groupthink, individuals mirror the dominant opinions of the group without critically reflecting on alternative viewpoints. This recursive loop of social conformity can stifle creativity and innovation, as individuals become trapped in the recursive mirroring of existing patterns rather than introducing new ideas. Groupthink highlights the importance of cognitive dislocation, the introduction of new perspectives or ideas that disrupt the recursive loop of conformity and allow for greater flexibility and innovation.

The brain's recursive processes also contribute to the formation and maintenance of social hierarchies. In hierarchical social structures, individuals recursively mirror the behaviours and expectations of those in positions of authority, reinforcing the power dynamics within the group. For example, in a classroom setting, students may mirror the behaviour and attitudes of their teacher, recursively adjusting their actions to align with the teacher's expectations. This recursive loop of social reinforcement helps maintain the structure of the hierarchy, as individuals conform to the roles and responsibilities assigned to them.

However, the recursive nature of social hierarchies can also be challenged through social dislocation, where individuals or groups introduce variations that disrupt the established patterns of behaviour and power. Social movements, for instance, often arise when individuals mirror the dissatisfaction or frustration of others, recursively amplifying these emotions until they reach a tipping point. At this point, the recursive loop of conformity is broken, and new social patterns, such as demands for equality or justice, emerge. This recursive process of social change highlights the brain's capacity to both reinforce and disrupt social patterns in response to evolving needs and challenges.

In parent-child relationships, recursive processes are central to the development of attachment and social learning. Children mirror the behaviours and emotions of their caregivers, recursively refining their understanding of social norms, language, and emotional regulation. This recursive mirroring is essential for the development of secure attachment, where children feel safe and supported in their interactions with their caregivers. Over time, children internalise these social patterns, recursively reflecting on their experiences to build a sense of self and navigate the social world.

The recursive nature of social learning is also evident in peer relationships, where children and adolescents mirror the behaviours, language, and attitudes of their friends. As they engage in social interactions, they recursively adjust their behaviour in response to social feedback, refining their understanding of group dynamics and social expectations. This recursive process is particularly important during adolescence, a period of significant social and emotional development. Peer relationships during this time are characterised by recursive cycles of social reinforcement, where individuals adjust their behaviour to fit in with their social group while simultaneously influencing the group's norms and values.

Cultural transmission, the process by which knowledge, traditions, and social norms are passed from one generation to the next, is another example of recursive social dynamics. Cultural transmission relies on the recursive mirroring of behaviours, language, and beliefs, where individuals observe and replicate the patterns of their culture. Over time, these cultural patterns are refined and adapted in response to new challenges and innovations, ensuring that cultural knowledge evolves while maintaining continuity with the past.

In modern society, digital technology has introduced new forms of recursive social interaction, where individuals mirror and refine patterns of behaviour and communication in virtual spaces. Social media platforms, for instance, create recursive loops of social validation and feedback, where individuals post content, receive likes and comments, and adjust their behaviour based on the responses of their online community. This recursive cycle of social feedback reinforces certain behaviours, shaping how individuals present themselves and interact with others in the digital world.

However, the recursive nature of digital interaction also raises challenges, particularly in the context of echo chambers and social polarisation. In echo chambers, individuals recursively mirror the opinions and beliefs of their online communities, reinforcing their existing views without exposure to alternative perspectives. This recursive loop of social reinforcement can lead to polarisation, where individuals become more entrenched in their beliefs and less open to dialogue with those who hold different views. This highlights the importance of dislocation in social dynamics, introducing new perspectives and breaking the recursive loop of conformity to promote greater understanding and cooperation.

In conflict resolution, recursive processes play a crucial role in helping individuals and groups navigate disagreements and find common ground. By mirroring the emotions and perspectives of the other party, individuals can recursively adjust their own approach, seeking to understand the underlying issues and find mutually beneficial solutions. This recursive loop of empathy and reflection allows for the resolution of conflicts in a way that fosters cooperation and strengthens social bonds.

In summary, human relationships and social dynamics are profoundly shaped by the brain's recursive processes of mirroring, reflection, and refinement. Whether in the context of interpersonal relationships, group dynamics, or cultural transmission, the brain's ability to recursively adjust and refine social patterns allows us to form deep emotional connections, navigate complex social environments, and contribute to the evolution of social structures. By understanding the recursive nature of social interaction, we can foster more meaningful relationships and create more adaptive, flexible social systems.

In the next chapter, we will explore how recursive processes shape cultural evolution, examining how societies mirror and refine their cultural patterns over time to adapt to new challenges and opportunities.

References

Bunge, M., 2003. *Emergence and Convergence: Qualitative Novelty and the Unity of Knowledge.* Toronto: University of Toronto Press.

Dennett, D.C., 1991. *Consciousness Explained.* Boston: Little, Brown and Company.

Friston, K., 2010. The free-energy principle: A unified brain theory? *Nature Reviews Neuroscience*, 11(2), pp.127–138.

Gigerenzer, G., 2008. Rationality for mortals: How people cope with uncertainty. Oxford: Oxford University Press.

Kuhn, T.S., 1977. *The Essential Tension.* Chicago: University of Chicago Press.

Lakatos, I., 1970. Falsification and the methodology of scientific research programmes. In: I. Lakatos and A. Musgrave, eds. *Criticism and the Growth of Knowledge*. Cambridge: Cambridge University Press, pp.91–196.

Marr, D., 1982. *Vision: A Computational Investigation into the Human Representation and Processing of Visual Information*. San Francisco: W.H. Freeman.

Mitchell, M., 2009. *Complexity: A Guided Tour*. Oxford: Oxford University Press.

Popper, K., 1963. *Conjectures and Refutations*. London: Routledge.

Simon, H.A., 1996. *The Sciences of the Artificial*. 3rd ed. Cambridge, MA: MIT Press.

Tononi, G., 2008. Consciousness as integrated information: A provisional manifesto. *Biological Bulletin*, 215(3), pp.216–242.

Chapter 19:

Recursive Processes in Cultural Evolution and Societal Change

Cultural evolution, much like biological evolution, is a dynamic process shaped by the recursive mirroring, adaptation, and refinement of social patterns across generations. Societies do not exist in a state of fixed traditions and beliefs; instead, they evolve continuously, reflecting on past experiences while adapting to new challenges and opportunities. Recursive processes are central to how cultural knowledge, norms, and values are transmitted, transformed, and innovated. In this chapter, we will explore how recursive processes drive cultural evolution, how societies mirror and refine their traditions, and how these processes contribute to societal change.

Cultural evolution begins with the transmission of social patterns from one generation to the next. This transmission is deeply recursive, as each generation mirrors the patterns of its predecessors, integrating the behaviours, language, beliefs, and customs of the past into their own lives. However, cultural transmission is not a passive process of replication. Each generation introduces dislocations and adaptations to these inherited patterns, reflecting the new challenges and opportunities that arise in the changing world.

For example, consider the oral storytelling traditions of indigenous cultures. These stories are passed down through generations, with each storyteller mirroring the narrative patterns of their ancestors. However, as these stories are told and retold, each generation introduces subtle variations or embellishments that reflect the contemporary social and cultural context. This recursive refinement ensures that the stories remain relevant and meaningful while preserving the core themes and values of the culture.

Language itself evolves recursively, as speakers introduce new words, phrases, and grammatical structures that reflect changing social and technological realities. The evolution of language mirrors the dynamic nature of culture, where linguistic patterns are mirrored and refined over time. For example, the rapid development of digital technology has introduced new vocabulary, such as "hashtag," "selfie," or "streaming", that reflects the changing ways in which we interact with the world. These linguistic innovations are mirrored by speakers and recursively integrated into the language, ensuring that it evolves to meet the needs of a constantly changing society.

Religious practices also evolve through recursive processes. While many religious traditions are rooted in ancient texts and rituals, these practices are continuously adapted to reflect the needs and values of contemporary society. For instance, many religious communities have mirrored and reflected on social movements, such as the civil rights movement or the growing emphasis on environmental stewardship, to refine their teachings and practices. This recursive reflection allows religious traditions to remain relevant and responsive to the ethical concerns of the present while maintaining continuity with their spiritual heritage.

The role of recursive processes in cultural evolution is perhaps most evident in the realm of art and architecture. Artistic movements often emerge as responses to the dominant cultural patterns of the time, introducing dislocations that challenge traditional forms and concepts. For example, the transition from Renaissance art, with its emphasis on symmetry and realism, to modernist and abstract art involved a recursive break from the established patterns of representation. Artists such as Picasso and Kandinsky mirrored the patterns of earlier artistic traditions but introduced radical dislocations that reflected the changing intellectual, technological, and social landscape of the 20th century.

Architecture follows a similar recursive trajectory, where architectural styles evolve in response to social, technological, and environmental changes. The transition from classical to modern architecture, and later to postmodern and sustainable architecture, reflects a recursive refinement of building design and urban planning. For example, the shift towards sustainable architecture mirrors the growing recognition of environmental challenges, introducing new patterns of design that prioritise energy efficiency, ecological harmony, and the integration of natural elements.

Cultural evolution is also shaped by the brain's ability to recursively reflect on ethical dilemmas and social justice issues. As societies grapple with questions of fairness, equality, and human rights, they mirror the patterns of past social movements and refine their approaches to meet the challenges of the present. For example, the feminist movement, the civil rights movement, and the LGBTQ+ rights movement all mirror earlier struggles for justice, but introduce new dislocations and adaptations to reflect contemporary understandings of gender, race, and identity.

The recursive nature of social change is particularly evident in the way societies respond to technological innovation. Each new technology introduces dislocations in cultural patterns, forcing societies to reflect on how these innovations will impact social norms, ethics, and behaviour. For example, the advent of the internet and social media has transformed how individuals communicate, access information, and form communities. These technologies have mirrored and disrupted traditional forms of media and communication, introducing new patterns of social interaction that are continuously being refined.

One of the most significant aspects of recursive cultural evolution is the tension between tradition and innovation. While traditions serve as the foundation of cultural continuity, they are constantly being re-evaluated and adapted in light of new knowledge, challenges, and perspectives. For example, in many societies, traditional gender roles and family structures have been mirrored and refined in response to changing social norms, leading to greater gender equality and more diverse family models. This recursive process ensures that cultural traditions remain flexible and responsive to the needs of contemporary society.

Political systems also evolve recursively, reflecting the ongoing dialogue between the state and its citizens. Democratic systems, in particular, rely on recursive processes of feedback and adaptation, where governments reflect on the needs and desires of the population and adjust policies accordingly. The process of elections, legislation, and public discourse creates a recursive loop where social and political patterns are mirrored, debated, and refined over time. Political systems that fail to engage in recursive reflection and adaptation risk stagnation or collapse, as they become unable to respond to the evolving needs of their societies.

Economic systems, too, are shaped by recursive processes. The evolution of capitalism, for example, has been characterised by cycles of expansion, crisis, and reform, where economic patterns are mirrored, disrupted, and refined in response to changing market conditions and social demands. The Great Depression, for instance, introduced dislocations in the global economic system, prompting a recursive re-evaluation of economic theories and practices. This led to the development of Keynesian economics and the creation of social safety nets, which transformed economic policy in the 20th century.

Similarly, the 2008 financial crisis mirrored past economic disruptions but introduced new challenges that required recursive reflection on global financial systems.

Education plays a central role in the recursive transmission and refinement of cultural knowledge. Pedagogical practices evolve as educators mirror the learning needs of their students and adapt their teaching methods to reflect new research in cognitive development and technology. The rise of online learning platforms, for example, reflects a recursive adaptation to the digital age, allowing for more flexible, accessible, and personalised forms of education. Similarly, the emphasis on critical thinking and problem-solving in modern education mirrors the increasing complexity of the world, where individuals must be able to adapt and innovate in response to rapidly changing circumstances.

Cultural evolution is not limited to large-scale social systems; it also operates on the level of everyday life and individual behaviour. The brain's recursive ability to mirror and refine social patterns allows individuals to navigate the complexities of modern society, where traditional norms may no longer apply. For instance, individuals today must constantly adjust their behaviour to reflect the diverse cultural, ethical, and technological landscapes they encounter in work, social media, and personal relationships. This recursive adaptation allows for greater flexibility and openness to new ideas, but it also introduces challenges in maintaining a coherent sense of identity and community.

The recursive nature of cultural evolution is particularly evident in the arts, where innovation often arises from the recursive reflection on past artistic movements and the introduction of dislocations that challenge established norms. For example, the punk rock movement of the 1970s mirrored the social and musical patterns of earlier rock and counterculture movements but introduced a raw, minimalist aesthetic that reflected the disillusionment and rebellion of its time. Similarly, contemporary street art mirrors traditional forms of public art but introduces new patterns of political and social commentary that reflect the concerns of urban communities.

In modern times, globalisation has introduced new recursive dynamics in cultural evolution, where societies mirror and integrate patterns from diverse cultures, leading to cultural hybridisation. This process is evident in the fusion of food traditions, music genres, and fashion trends, where elements from different cultures are mirrored, combined, and refined to create new, hybrid forms of expression. Globalisation also creates dislocations, as traditional cultures confront the influences of global media, technology, and economic systems, leading to both the preservation and transformation of cultural identities.

In summary, cultural evolution is a deeply recursive process, where social patterns are mirrored, refined, and adapted across generations.

Whether in the transmission of traditions, the evolution of language, or the rise of social movements, recursive processes ensure that cultures remain flexible, responsive, and innovative. By understanding the recursive nature of cultural evolution, we can better appreciate how societies navigate the tension between tradition and innovation, ensuring that cultural knowledge is preserved while remaining open to transformation.

In the next chapter, we will explore how recursive processes shape global systems, examining how international relations, environmental policies, and economic systems are interconnected through recursive feedback loops that influence global stability and change.

References

Bransford, J.D., Brown, A.L. and Cocking, R.R., 2000. *How People Learn: Brain, Mind, Experience, and School*. Washington, DC: National Academy Press.

Hattie, J., 2009. *Visible Learning: A Synthesis of Over 800 Meta-Analyses Relating to Achievement*. London: Routledge.

Kolb, D.A., 1984. *Experiential Learning: Experience as the Source of Learning and Development*. Englewood Cliffs, NJ: Prentice-Hall.

Mayer, R.E., 2009. *Multimedia Learning*. 2nd ed. Cambridge: Cambridge University Press.

Piaget, J., 1973. *To Understand is to Invent: The Future of Education*. New York: Grossman Publishers.

Schunk, D.H., 2012. *Learning Theories: An Educational Perspective*. 6th ed. Boston: Pearson.

Sfard, A., 1998. On two metaphors for learning and the dangers of choosing just one. *Educational Researcher*, 27(2), pp.4–13.

Siemens, G., 2005. Connectivism: A learning theory for the digital age. *International Journal of Instructional Technology and Distance Learning*, 2(1), pp.3–10.

Sweller, J., 1988. Cognitive load during problem solving: Effects on learning. *Cognitive Science*, 12(2), pp.257–285.

Vygotsky, L.S., 1978. *Mind in Society*. Cambridge, MA: Harvard University Press.

Wenger, E., 1998. *Communities of Practice: Learning, Meaning, and Identity*. Cambridge: Cambridge University Press.

Chapter 20:

Recursive Processes in Global Systems and Interconnected Societies

In an increasingly interconnected world, global systems, from international relations to environmental policies and economic frameworks, are shaped by the recursive feedback loops that connect nations, societies, and ecosystems. These recursive processes ensure that actions taken at one level of the global system have ripple effects across other areas, influencing both local and global stability. In this chapter, we will explore how recursive processes shape global governance, economic systems, and environmental sustainability, and how these interconnections lead to both cooperation and conflict in the modern world.

Global systems, by their nature, are interdependent, meaning that no country or region operates in isolation. The actions of one country, whether in the political, economic, or environmental sphere, are mirrored and reflected across other nations, creating a recursive loop of cause and effect. This recursive interconnection means that decisions made by global actors, such as governments, multinational corporations, and international organisations, can have profound impacts on global stability and development.

One of the most visible examples of recursive processes in global systems is the dynamic of international trade. The global economy is an intricate network of interconnected markets, where goods, services, and capital flow between nations. When one country makes a policy change, such as imposing tariffs or subsidies, this action is mirrored by trading partners and competitors, who adjust their own policies in response. This recursive loop of action and reaction leads to shifts in the global supply chain, affecting everything from commodity prices to employment and growth rates in different regions. As countries recursively mirror each other's economic decisions, they create patterns of global trade that are continuously refined and adjusted in response to changing market conditions.

The financial system is another area where recursive processes play a critical role. Global financial markets are deeply interconnected, meaning that shocks in one market can quickly spread across the world. For example, the 2008 financial crisis began with the collapse of the housing market in the United States but quickly spread to global financial institutions, triggering recessions in countries around the world. The recursive nature of global finance means that economic policies and decisions are mirrored across different markets, amplifying both risks and opportunities. Central banks, governments, and financial institutions must recursively adjust their policies in response to these global dynamics, using tools such as interest rate changes, quantitative easing, and regulatory reforms to stabilise the system.

In the realm of international relations, recursive processes govern the way countries interact, forming alliances, negotiating treaties, and resolving conflicts. The global system of diplomacy is built on recursive feedback loops, where nations mirror each other's actions and adjust their strategies in response. For instance, when one country increases its military presence in a region, neighbouring countries often mirror this action by enhancing their own defence capabilities, leading to a recursive arms race. Similarly, when countries engage in diplomatic negotiations or trade agreements, their actions are mirrored and refined through a process of give and take, where each party adjusts its position based on the actions of the other.

Multilateral organisations, such as the United Nations (UN), the World Trade Organization (WTO), and the International Monetary Fund (IMF), play a crucial role in managing these recursive processes at the global level. These organisations provide forums for nations to mirror and reflect on each other's policies, facilitating cooperation and resolving disputes through recursive negotiation and consensus-building. The Paris Agreement on climate change, for example, is the result of a recursive process of negotiation, where countries mirrored each other's commitments to reducing carbon emissions and refined their policies over time to address global environmental challenges.

Environmental sustainability is one of the most pressing global issues that illustrates the recursive interconnection between local actions and global outcomes. Ecosystems are highly recursive, where changes in one part of the system, such as deforestation, pollution, or overfishing, can lead to cascading effects across the entire system. Human activity has disrupted many of these natural recursive processes, leading to climate change, biodiversity loss, and ecosystem degradation. For example, the burning of fossil fuels in one region contributes to the global buildup of greenhouse gases, which in turn affects weather patterns, sea levels, and agricultural productivity worldwide. The recursive nature of ecological systems means that environmental policies must account for the interconnectedness of local actions and their global consequences.

Climate change itself is a recursive challenge, where the accumulation of greenhouse gases creates feedback loops that exacerbate the problem. For instance, as the polar ice caps melt, less sunlight is reflected back into space, leading to further warming, a process known as the albedo effect. Similarly, as forests are destroyed, the carbon stored in trees is released into the atmosphere, accelerating the process of global warming. These recursive feedback loops highlight the urgency of addressing climate change at both the local and global levels. International efforts, such as the Paris

Agreement, aim to break these negative feedback loops by encouraging nations to adopt policies that reduce emissions and promote sustainability.

The recursive nature of global systems also shapes technological innovation and the digital economy. The development of new technologies, such as artificial intelligence (AI), blockchain, and renewable energy, creates recursive feedback loops, where technological advances drive changes in industries, markets, and social behaviour. For example, the rise of AI and automation has mirrored the recursive refinement of labour markets, where certain jobs are displaced by technology while new opportunities emerge in other sectors. Similarly, the adoption of renewable energy technologies has led to recursive shifts in global energy markets, with countries adjusting their energy policies in response to both technological innovation and environmental goals.

The digital economy is another area where recursive processes are highly visible. Digital platforms, such as social media, e-commerce sites, and online marketplaces, create recursive feedback loops between consumers, businesses, and data analytics. For example, when a consumer makes a purchase online, their behaviour is mirrored by algorithms that track their preferences and refine future recommendations. This recursive loop between user data and personalised experiences drives the growth of the digital economy, as platforms continuously refine their algorithms to provide more targeted services and products.

Global health is another domain where recursive processes are critical. The spread of infectious diseases, such as the COVID-19 pandemic, illustrates how local health crises can quickly become global challenges through recursive patterns of transmission. As countries mirrored each other's public health responses, such as lockdowns, travel restrictions, and vaccination campaigns, they refined their strategies based on emerging data and feedback. This recursive adaptation was essential for managing the pandemic on a global scale, as nations learned from each other's successes and failures, refining their approaches in real time.

The pandemic also highlighted the recursive nature of global supply chains, where disruptions in one region affected the availability of goods and services worldwide. For example, the initial outbreak of COVID-19 in China led to factory shutdowns, which disrupted the global production of electronics, pharmaceuticals, and other essential goods. These supply chain disruptions were mirrored across the world, creating a recursive loop of shortages and delays. The global response to these challenges involved recursive adjustments to supply chain management, as companies and governments refined their strategies to ensure resilience in the face of future disruptions.

Global inequality is another area where recursive processes are evident. Economic policies, trade agreements, and technological innovations often mirror the existing patterns of wealth distribution, reinforcing the divide between wealthy and developing nations. However, recursive dislocations, such as social movements, policy reforms, and technological breakthroughs, can introduce shifts in these patterns, creating opportunities for greater equity and inclusion. For example, the rise of mobile banking and digital payment platforms has allowed millions of people in developing countries to access financial services, disrupting traditional economic patterns and creating new pathways for economic empowerment.

The recursive processes that shape global systems also highlight the importance of global governance and collaborative decision-making. In an interconnected world, no single country or organisation can address global challenges alone. The complexity of global systems requires a recursive approach to governance, where nations, institutions, and civil society work together to mirror and refine their policies in response to emerging challenges. This recursive collaboration is essential for managing issues such as climate change, economic inequality, technological innovation, and global health, all of which require coordinated efforts across borders.

In summary, recursive processes are at the heart of global systems, shaping everything from international relations and economic policy to environmental sustainability and technological innovation. The interconnectedness of these systems means that local actions have global consequences, creating recursive feedback loops that influence global stability and change. By understanding the recursive nature of global systems, we can develop more effective strategies for addressing the complex challenges of the 21st century, ensuring that global cooperation, innovation, and sustainability are at the forefront of our efforts.

In the next chapter, we will conclude by synthesising the key insights from our exploration of recursive processes and examining how the recursive nature of the brain, society, and the global system informs our understanding of complexity, adaptation, and human progress.

References

Ackoff, R.L., 1971. Towards a system of systems concepts. *Management Science*, 17(11), pp.661–671.

Bateson, G., 1972. *Steps to an Ecology of Mind*. Chicago: University of Chicago Press.

Checkland, P., 1981. *Systems Thinking, Systems Practice*. Chichester: Wiley.

Gell-Mann, M., 1994. *The Quark and the Jaguar: Adventures in the Simple and the Complex*. London: Abacus.

Holland, J.H., 1998. *Emergence: From Chaos to Order*. Oxford: Oxford University Press.

Meadows, D.H., 2008. *Thinking in Systems: A Primer*. White River Junction, VT: Chelsea Green Publishing.

Morin, E., 2008. *On Complexity*. Cresskill, NJ: Hampton Press.

Newell, A., 1990. *Unified Theories of Cognition*. Cambridge, MA: Harvard University Press.

Polanyi, M., 1966. *The Tacit Dimension*. London: Routledge.

Resnick, M., 1994. *Turtles, Termites, and Traffic Jams*. Cambridge, MA: MIT Press.

Simon, H.A., 1996. *The Sciences of the Artificial*. 3rd ed. Cambridge, MA: MIT Press.

von Bertalanffy, L., 1968. *General System Theory*. New York: George Braziller.

Chapter 21:

Synthesising Recursive Patterns: Complexity, Adaptation, and Human Progress

As we have explored throughout this book, recursive processes are fundamental to the functioning of the human brain, individual creativity, social interactions, cultural evolution, and global systems. These recursive loops of mirroring, reflection, and refinement enable us to adapt, innovate, and thrive in an ever-changing world. In this concluding chapter, we will synthesise the key insights from our examination of recursive processes and consider how they inform our understanding of complexity, adaptation, and human progress.

At the core of recursion is the brain's remarkable ability to mirror patterns from its internal and external environments, introducing slight variations or dislocations to refine its understanding and response to new information. Whether in the cognitive processes that underlie memory, creativity, or consciousness, or in the more complex behaviours associated with social interactions and learning, the recursive loops of reflection and refinement allow us to make sense of the world and adjust to its complexities.

One of the most important themes to emerge from our exploration is the role of recursion in managing complexity. Both the brain and society are complex, adaptive systems that rely on recursive processes to navigate and respond to the vast amounts of information they encounter. The human brain, with its billions of neurons and countless synaptic connections, relies on recursive loops of pattern recognition and reflection to organise and prioritise information. Similarly, societies and global systems manage the complexity of economic, political, and social forces through recursive feedback loops that allow them to remain flexible and adaptive.

Complexity theory, which originated in fields such as physics and mathematics, provides a useful framework for understanding how recursive processes manage complexity. In complex systems, small variations in initial conditions can lead to significant outcomes through recursive feedback loops, a phenomenon often described as the butterfly effect. The ability of both biological and social systems to evolve and adapt over time is rooted in these recursive dynamics, where systems continually mirror and refine patterns in response to changing conditions. The brain's recursive processes enable it to handle the complexity of thought and emotion, while recursive feedback loops in society allow cultures, economies, and political systems to evolve and adapt to new challenges.

In this sense, adaptation, whether at the level of the individual, society, or the global system, relies on recursion as a mechanism for learning and growth. Biological evolution, as discussed in earlier chapters, is itself a recursive process where genetic patterns are replicated, with occasional dislocations in the form of mutations leading to new adaptations. Similarly, in the cognitive realm, recursion enables creative problem-solving, allowing individuals to reflect on familiar patterns and introduce novel variations that lead to breakthroughs in science, art, and technology. This recursive capacity for adaptation is also what enables societies to remain resilient in the face of crises, such as economic downturns, environmental degradation, or pandemics.

Human progress, both individually and collectively, can be seen as the outcome of recursive processes of reflection and refinement. As individuals, we recursively refine our understanding of ourselves and the world around us, learning from our experiences and adapting our behaviour to meet new goals. At the societal level, cultures and political systems mirror past patterns while introducing dislocations that allow for social innovation and reform. Scientific progress, too, is the result of recursive cycles of hypothesis, experimentation, and refinement, where new knowledge is built on the recursive reflection of previous discoveries.

However, as we have seen, recursion is not without its challenges. While recursive loops allow for learning and adaptation, they can also lead to negative patterns of behaviour, such as groupthink, social polarisation, or environmental degradation. In the absence of dislocation, the introduction of new perspectives or changes that disrupt harmful recursive patterns, systems can become rigid and resistant to change. The recursive loops of social media, for example, can reinforce echo chambers where individuals mirror and amplify their existing beliefs without exposure to alternative viewpoints, leading to greater division and less understanding across societal lines.

This points to the importance of cognitive dislocation in fostering flexibility and resilience. In both individual and collective systems, the introduction of new ideas, perspectives, or challenges is essential for breaking negative recursive loops and allowing for creative adaptation. This is why diversity of thought, critical reflection, and open dialogue are so important in ensuring that recursive processes lead to positive outcomes. In science, art, politics, and everyday life, dislocations introduce the necessary variations that prevent stagnation and promote growth.

The role of recursive processes in global systems has become increasingly apparent in the 21st century, as nations and societies become more interconnected. Globalisation, with its recursive feedback loops between local actions and global consequences, highlights the complexity of managing economic, environmental, and political systems on a planetary scale. The recursive nature of climate change, for example, shows how local activities, such as deforestation or carbon emissions, can create global feedback loops that amplify environmental degradation. Addressing these challenges requires a recursive approach to global governance, where nations mirror each other's policies, reflect on their impacts, and refine their strategies in response to global goals.

Technological innovation, too, is shaped by recursive processes, where new technologies are built on the recursive refinement of previous inventions. The development of artificial intelligence (AI), for instance, mirrors the brain's recursive processes of learning and adaptation, with machine learning algorithms continuously refining their performance based on feedback from data. As we move further into the age of automation and digital technologies, the recursive interplay between human intelligence and AI systems will likely shape the future of work, education, and social interaction.

In the arts, recursion is a central theme in the creative process, where artists, musicians, and writers mirror and refine patterns to create new forms of expression. Pablo Picasso's use of recursive motifs in his works, James Joyce's recursive narrative structures, and Beethoven's evolving symphonic themes all exemplify how recursion drives artistic innovation. These recursive patterns allow for the layering of complexity and meaning, creating works that resonate with audiences across time and culture.

Ultimately, the recursive processes we have explored throughout this book illustrate the interconnectedness of the brain, society, and the global system. Whether through the brain's capacity for self-awareness and introspection, the evolution of social dynamics and cultural identities, or the recursive feedback loops that shape global governance and environmental policy, recursion is a fundamental principle of life and progress. It allows systems to remain adaptable, flexible, and responsive to change, ensuring that human beings, and the societies they create, can navigate the complexities of an evolving world.

In conclusion, recursion provides us with a powerful lens for understanding the nature of human cognition, creativity, and societal change. By recognising the recursive patterns that shape our thoughts, behaviours, and institutions, we can better appreciate the mechanisms that drive adaptation, learning, and innovation. More importantly, by embracing the recursive nature of progress, we can cultivate the flexibility and resilience needed to address the challenges of the future, ensuring that we continue to evolve and thrive as individuals, societies, and as a global community.

References

Arthur, W.B., 2009. *The Nature of Technology: What It Is and How It Evolves*. New York: Free Press.

Basalla, G., 1988. *The Evolution of Technology*. Cambridge: Cambridge University Press.

Christensen, C.M., 1997. *The Innovator's Dilemma*. Boston: Harvard Business School Press.

Fagerberg, J., Mowery, D.C. and Nelson, R.R., 2005. *The Oxford Handbook of Innovation*. Oxford: Oxford University Press.

Hughes, T.P., 1987. The evolution of large technological systems. In: W.E. Bijker, T.P. Hughes and T.J. Pinch, eds. *The Social Construction of Technological Systems*. Cambridge, MA: MIT Press, pp.51–82.

Mokyr, J., 1990. *The Lever of Riches: Technological Creativity and Economic Progress*. Oxford: Oxford University Press.

Nelson, R.R. and Winter, S.G., 1982. *An Evolutionary Theory of Economic Change*. Cambridge, MA: Harvard University Press.

Rogers, E.M., 2003. *Diffusion of Innovations*. 5th ed. New York: Free Press.

Schumpeter, J.A., 1942. *Capitalism, Socialism and Democracy*. New York: Harper and Brothers.

Utterback, J.M., 1994. *Mastering the Dynamics of Innovation*. Boston: Harvard Business School Press.

von Hippel, E., 2005. *Democratizing Innovation*. Cambridge, MA: MIT Press.

Bibliography

Neuroscience and Cognition:

- Damasio, A. (1999). *The Feeling of What Happens: Body and Emotion in the Making of Consciousness*. Harcourt Brace.
- Edelman, G. M. (1989). *The Remembered Present: A Biological Theory of Consciousness*. Basic Books.
- Friston, K. J. (2010). "The Free-Energy Principle: A Unified Brain Theory?" *Nature Reviews Neuroscience*, 11, 127–138.
- LeDoux, J. (1998). *The Emotional Brain: The Mysterious Underpinnings of Emotional Life*. Simon & Schuster.
- McGilchrist, I. (2019). *The Master and His Emissary: The Divided Brain and the Making of the Western World*. Yale University Press.
- Ramachandran, V. S. (2011). *The Tell-Tale Brain: A Neuroscientist's Quest for What Makes Us Human*. W. W. Norton & Company.
- Squire, L. R., & Kandel, E. R. (2009). *Memory: From Mind to Molecules*. Roberts & Company Publishers.
- Tononi, G., & Koch, C. (2015). "Consciousness: Here, There and Everywhere?" *Philosophical Transactions of the Royal Society B*, 370(1668), 20140167.

- Zeki, S. (1999). *Inner Vision: An Exploration of Art and the Brain*. Oxford University Press.

Psychology and Learning:

- Baddeley, A. (2007). *Working Memory, Thought, and Action*. Oxford University Press.
- Bandura, A. (1977). *Social Learning Theory*. Prentice-Hall.
- Piaget, J. (1954). *The Construction of Reality in the Child*. Basic Books.
- Vygotsky, L. S. (1978). *Mind in Society: The Development of Higher Psychological Processes*. Harvard University Press.
- Schön, D. A. (1983). *The Reflective Practitioner: How Professionals Think in Action*. Basic Books.
- Siegler, R. S. (2005). "Children's Learning." *American Psychologist*, 60(8), 769–778.
- Ericsson, K. A., & Pool, R. (2016). *Peak: Secrets from the New Science of Expertise*. Mariner Books.

Philosophy and Systems Theory:

- Hofstadter, D. R. (1979). *Gödel, Escher, Bach: An Eternal Golden Braid*. Basic Books.
- Maturana, H. R., & Varela, F. J. (1980). *Autopoiesis and Cognition: The Realization of the Living*. Reidel.

- Morin, E. (2008). *On Complexity*. Hampton Press.
- Nagel, T. (1974). "What Is It Like to Be a Bat?" *The Philosophical Review*, 83(4), 435–450.
- Popper, K. R. (1959). *The Logic of Scientific Discovery*. Hutchinson.
- Searle, J. R. (1992). *The Rediscovery of the Mind*. MIT Press.
- Thompson, E. (2007). *Mind in Life: Biology, Phenomenology, and the Sciences of Mind*. Harvard University Press.
- Wheeler, J. A. (1990). "Information, Physics, Quantum: The Search for Links." In Zurek, W. H. (Ed.), *Complexity, Entropy, and the Physics of Information*. Addison-Wesley.

Art, Creativity, and Culture:

- Arnheim, R. (1954). *Art and Visual Perception: A Psychology of the Creative Eye*. University of California Press.
- Deleuze, G., & Guattari, F. (1987). *A Thousand Plateaus: Capitalism and Schizophrenia*. University of Minnesota Press.
- Picasso, P. (1946). *The Creative Process in Art*. HarperCollins.
- Rosenberg, H. (1959). *The Tradition of the New*. Horizon Press.

- Tolstoy, L. (1897). *What Is Art?* Penguin Classics.
- Turner, V. (1969). *The Ritual Process: Structure and Anti-Structure.* Aldine.

Social Theory and Group Dynamics:

- Bauman, Z. (2000). *Liquid Modernity.* Polity Press.
- Bourdieu, P. (1990). *The Logic of Practice.* Stanford University Press.
- Durkheim, E. (1912). *The Elementary Forms of Religious Life.* Free Press.
- Goffman, E. (1959). *The Presentation of Self in Everyday Life.* Anchor Books.
- Habermas, J. (1984). *The Theory of Communicative Action, Vol. 1: Reason and the Rationalization of Society.* Beacon Press.
- Luhmann, N. (1995). *Social Systems.* Stanford University Press.
- Tajfel, H., & Turner, J. C. (1979). "An Integrative Theory of Intergroup Conflict." In Austin, W. G., & Worchel, S. (Eds.), *The Social Psychology of Intergroup Relations.* Brooks/Cole.

Cultural Evolution and Global Systems:

- Diamond, J. (1997). *Guns, Germs, and Steel: The Fates of Human Societies*. W. W. Norton & Company.
- Harari, Y. N. (2015). *Sapiens: A Brief History of Humankind*. Harper.
- Huntington, S. P. (1996). *The Clash of Civilizations and the Remaking of World Order*. Simon & Schuster.
- Latour, B. (2005). *Reassembling the Social: An Introduction to Actor-Network Theory*. Oxford University Press.
- Margulis, L. (1998). *Symbiotic Planet: A New Look at Evolution*. Basic Books.
- Polanyi, K. (1944). *The Great Transformation: The Political and Economic Origins of Our Time*. Beacon Press.
- Wallerstein, I. (2004). *World-Systems Analysis: An Introduction*. Duke University Press.

Complexity, Adaptation, and Evolution:

- Kauffman, S. A. (1995). *At Home in the Universe: The Search for the Laws of Self-Organization and Complexity*. Oxford University Press.
- Mayr, E. (2001). *What Evolution Is*. Basic Books.
- Prigogine, I., & Stengers, I. (1984). *Order Out of Chaos: Man's New Dialogue with Nature*. Bantam Books.
- Simon, H. A. (1969). *The Sciences of the Artificial*. MIT Press.

- Wilson, E. O. (1975). *Sociobiology: The New Synthesis.* Harvard University Press.
- Wright, R. (2000). *Nonzero: The Logic of Human Destiny.* Vintage Books.

Technology, AI, and the Digital World:

- Brynjolfsson, E., & McAfee, A. (2014). *The Second Machine Age: Work, Progress, and Prosperity in a Time of Brilliant Technologies.* W. W. Norton & Company.
- Kurzweil, R. (2005). *The Singularity Is Near: When Humans Transcend Biology.* Viking.
- Lanier, J. (2010). *You Are Not a Gadget: A Manifesto.* Knopf.
- Moravec, H. (1999). *Robot: Mere Machine to Transcendent Mind.* Oxford University Press.
- Tegmark, M. (2017). *Life 3.0: Being Human in the Age of Artificial Intelligence.* Knopf.
- Zuboff, S. (2019). *The Age of Surveillance Capitalism: The Fight for a Human Future at the New Frontier of Power.* PublicAffairs.

Further Reading

Arnheim, R., 1954. *Art and Visual Perception: A Psychology of the Creative Eye*. Berkeley: University of California Press.

Baddeley, A., 2007. *Working Memory, Thought, and Action*. Oxford: Oxford University Press.

Bandura, A., 1977. *Social Learning Theory*. Englewood Cliffs, NJ: Prentice-Hall.

Bauman, Z., 2000. *Liquid Modernity*. Cambridge: Polity Press.

Bourdieu, P., 1990. *The Logic of Practice*. Stanford, CA: Stanford University Press.

Brynjolfsson, E. & McAfee, A., 2014. *The Second Machine Age: Work, Progress, and Prosperity in a Time of Brilliant Technologies*. New York: W. W. Norton & Company.

Damasio, A., 1999. *The Feeling of What Happens: Body and Emotion in the Making of Consciousness*. New York: Harcourt Brace.

Deleuze, G. & Guattari, F., 1987. *A Thousand Plateaus: Capitalism and Schizophrenia*. Minneapolis: University of Minnesota Press.

Diamond, J., 1997. *Guns, Germs, and Steel: The Fates of Human Societies*. New York: W. W. Norton & Company.

Durkheim, E., 1912. *The Elementary Forms of Religious Life*. New York: Free Press.

Edelman, G. M., 1989. *The Remembered Present: A Biological Theory of Consciousness*. New York: Basic Books.

Ericsson, K. A. & Pool, R., 2016. *Peak: Secrets from the New Science of Expertise*. New York: Mariner Books.

Friston, K. J., 2010. The free-energy principle: A unified brain theory? *Nature Reviews Neuroscience*, 11, pp. 127–138.

Goffman, E., 1959. *The Presentation of Self in Everyday Life*. New York: Anchor Books.

Habermas, J., 1984. *The Theory of Communicative Action, Vol. 1: Reason and the Rationalization of Society*. Boston: Beacon Press.

Harari, Y. N., 2015. *Sapiens: A Brief History of Humankind*. New York: Harper.

Hofstadter, D. R., 1979. *Gödel, Escher, Bach: An Eternal Golden Braid*. New York: Basic Books.

Hume, D., 1978. *A Treatise of Human Nature*. Oxford: Clarendon Press.

Huntington, S. P., 1996. *The Clash of Civilizations and the Remaking of World Order*. New York: Simon & Schuster.

Kauffman, S. A., 1995. *At Home in the Universe: The Search for the Laws of Self-Organization and Complexity*. New York: Oxford University Press.

Kurzweil, R., 2005. *The Singularity Is Near: When Humans Transcend Biology*. New York: Viking.

Lanier, J., 2010. *You Are Not a Gadget: A Manifesto*. New York: Knopf.

Latour, B., 2005. *Reassembling the Social: An Introduction to Actor-Network Theory*. Oxford: Oxford University Press.

LeDoux, J., 1998. *The Emotional Brain: The Mysterious Underpinnings of Emotional Life*. New York: Simon & Schuster.

Luhmann, N., 1995. *Social Systems*. Stanford, CA: Stanford University Press.

Margulis, L., 1998. *Symbiotic Planet: A New Look at Evolution*. New York: Basic Books.

Maturana, H. R. & Varela, F. J., 1980. *Autopoiesis and Cognition: The Realization of the Living*. Dordrecht: Reidel.

Mayr, E., 2001. *What Evolution Is*. New York: Basic Books.

McGilchrist, I., 2019. *The Master and His Emissary: The Divided Brain and the Making of the Western World*. New Haven, CT: Yale University Press.

Moravec, H., 1999. *Robot: Mere Machine to Transcendent Mind*. New York: Oxford University Press.

Morin, E., 2008. *On Complexity*. Creskill, NJ: Hampton Press.

Nagel, T., 1974. What is it like to be a bat? *The Philosophical Review*, 83(4), pp. 435–450.

Picasso, P., 1946. *The Creative Process in Art*. New York: HarperCollins.

Piaget, J., 1954. *The Construction of Reality in the Child*. New York: Basic Books.

Polanyi, K., 1944. *The Great Transformation: The Political and Economic Origins of Our Time*. Boston: Beacon Press.

Popper, K. R., 1959. *The Logic of Scientific Discovery*. London: Hutchinson.

Prigogine, I. & Stengers, I., 1984. *Order Out of Chaos: Man's New Dialogue with Nature*. New York: Bantam Books.

Ramachandran, V. S., 2011. *The Tell-Tale Brain: A Neuroscientist's Quest for What Makes Us Human*. New York: W. W. Norton & Company.

Rosenberg, H., 1959. *The Tradition of the New*. New York: Horizon Press.

Searle, J. R., 1992. *The Rediscovery of the Mind*. Cambridge, MA: MIT Press.

Schön, D. A., 1983. *The Reflective Practitioner: How Professionals Think in Action*. New York: Basic Books.

Searle, J. R., 1992. *The Rediscovery of the Mind*. Cambridge, MA: MIT Press.

Siegler, R. S., 2005. Children's learning. *American Psychologist*, 60(8), pp. 769–778.

Simon, H. A., 1969. *The Sciences of the Artificial*. Cambridge, MA: MIT Press.

Squire, L. R. & Kandel, E. R., 2009. *Memory: From Mind to Molecules*. Englewood, CO: Roberts & Company Publishers.

Tajfel, H. & Turner, J. C., 1979. An integrative theory of intergroup conflict. In: W. G. Austin & S. Worchel, eds. *The Social Psychology of Intergroup Relations*. Monterey, CA: Brooks/Cole, pp. 33-47.

Tegmark, M., 2017. *Life 3.0: Being Human in the Age of Artificial Intelligence*. New York: Knopf.

Thompson, E., 2007. *Mind in Life: Biology, Phenomenology, and the Sciences of Mind*. Cambridge, MA: Harvard University Press.

Tolstoy, L., 1897. *What Is Art?* London: Penguin Classics.

Tononi, G. & Koch, C., 2015. Consciousness: Here, there, and everywhere? *Philosophical Transactions of the Royal Society B*, 370(1668), p. 20140167.

Turner, V., 1969. *The Ritual Process: Structure and Anti-Structure*. Chicago: Aldine.

Vygotsky, L. S., 1978. *Mind in Society: The Development of Higher Psychological Processes*. Cambridge, MA: Harvard University Press.

Wallerstein, I., 2004. *World-Systems Analysis: An Introduction*. Durham, NC: Duke University Press.

Wilson, E. O., 1975. *Sociobiology: The New Synthesis*. Cambridge, MA: Harvard University Press.

Wright, R., 2000. *Nonzero: The Logic of Human Destiny*. New York: Vintage Books.

Zeki, S., 1999. *Inner Vision: An Exploration of Art and the Brain*. Oxford: Oxford University Press.

Zuboff, S., 2019. *The Age of Surveillance Capitalism: The Fight for a Human Future at the New Frontier of Power*. New York: PublicAffairs.

www.ingramcontent.com/pod-product-compliance
Lightning Source LLC
Chambersburg PA
CBHW020646220526
45464CB00001B/313